SPEC!AL

ARTIST COLLABORATIONS ON PACKAGING DESIGN

EDiTI0N

viction:ary

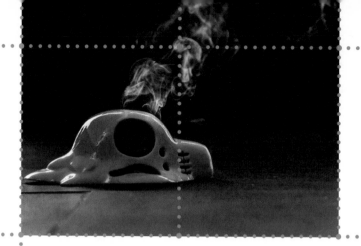

BY
CASE
STUDYO

We live in a world of big industry and mass production. The industry's conveyor belt never seems to stop spewing out products. These objects have a function but often seem to lack a soul.

The overload of consumer society fills the world, but leaves our hearts empty. All that abundance can't provide what the consumer actually yearns for; a little bit of love and consideration. We don't merely want to be catered to — we want to be cared for.

No matter how cold the industrialised world may seem, a heart-warmingly growing number of people have made it their mission to reintroduce the love for craft in the production process. A counterculture of thoughtful companies, product designers and agencies specialising in creative collaborations is on the rise, dedicated to soulful output of the highest quality. Society's need for decompression translates itself in these companies' production process. Restraints of shape, budget and material choices are considered and often present new possibilities. Things are given thought and every aspect of the product is treated with care to make it shine in its own right. Packaging can become a silkscreened artwork or a unique way to display the product.

Case Studyo's 'Artist Series' is an homage to these soldiers of love, the craftsmen and designers that give merit to the smaller things and strive to create a legacy of objects that capture the time we live in.

" [WE] STRIVE TO CREATE A LEGACY OF OBJECTS THAT CAPTURE THE TIME WE LIVE IN. A TIME OF ABOLISHING BOUNDARIES — BOUNDARIES BETWEEN HIGH AND LOW, ART AND DESIGN OR EVEN FORM AND FUNCTION. "

A time of abolishing boundaries — boundaries between high and low, art and design or even form and function.

At Case Studyo we cherish our small productions and collaborations with artists. The curation of our artists is guided by our personal taste and gut feeling. We like to work with new talent as well as established artists. There are no creative boundaries. Artists can come from a background in graphic design, graffiti or contemporary art.

The idea of infusing meticulous care into the production process and thereby elevating the product to become a piece of art is the heart of our philosophy. Some of our series specifically focus on the duality between an art piece and a functional object.

We believe a limitless mindset is the key. We don't consider the lines to be blurred. We ignore them all together and strive for a perfect marriage between artistry and production. We are proud to be included in this publication documenting the forefront of this 'artisanal revolution'.

> " SPECIAL EDITION PRODUCT DESIGN IS A FIELD WHERE NO RULES OR FORMS APPLIES. FROM ITS SELECTION OF MATERIALS, EXCEPTIONAL PRODUCTION PROCESS, TO FINAL PACKAGING, IMAGINATION IS SET OUT TO ATTAIN ITS OWN STATUS. "

Special edition product design is a field where no rules or forms applies. From its selection of materials, exceptional production process, to final packaging, imagination is set out to attain its own status. Made specifically to boost brand images, designers and brands are challenged more than ever to surprise and impress potential collectors while bringing new energy into all participating parties. Rare designer collaborations or crossover designs are just some of the common ways for fashion labels, luxury brands and food manufacturers to create new demand for a known product. Limited edition product used to have a connotation of being unaffordable, and out of reach for the general population. However, that has changed; not just exclusive to luxury goods, necessity goods are on the rise in boosting brand identity, gain loyal clientele and making products affordable.

The motive for creating a special edition design can be number of things. For some it is an opportunity to "treat" their customer, by adding delicacy to a product. The creation does not necessarily have to be a new product, in fact, numerous entries featured in the book are classic products with just a brand new fashionable outfit. It not only boosts desirability to these products but also solidifies the brand's relationship with loyal customers. Other brands commission unique designs to share the joy of a special occasion, as varied as anniversaries or the introduction of seasonal products. And there is merging of artistic talent for a good cause; for instance, gathering public attention to show support towards a charity foundation and raise awareness of an important political and social issue in the society. Most commonly brands initiate

a project to self-promote or crossover collaborations with other brands to reach out to new audience. This often involves giving away gifts to customers, potential clients and business associates. Furthermore, special edition packaging allow brands to play with their image and add originality to products when the regular packaging can't be changed that often. Take *Campbell's Warhol Soup Cans* (p.236) by DDW for an example, ever since the leading pop art figure Andy Warhol created the Campbell Soup painting back in 1962, the classic Campbell soup label has become a cultural image. To celebrate 50th anniversary of Warhol's creation and to pay tribute to the artist, The Campbell Soup Company launched a special product with the classic soup dressed in new packages. This project is not only refreshing and relatable, but also adds authenticity to the original product. Even a non-collector would be tempted to buy the product.

Amongst the huge volume of quality work existing in the field, our priorities were on finding product design with distinct characteristics. Together, entries can be categorised into three main sections, they are *Play x Imitation*, *Form x Concept*, and *Vision x Illustration*. The first chapter features playful designs that encourage direct interaction with the audience. Take a look at *All Fit to Your Game* by adidas (p.076), the project is the sports brand's Fall/Winter athletes' package turned into a custom board

game including elements like cards and dice. Playfully listing training and product information like a game instruction, the message behind the game set is that they wish athletes to be on top of their game. *Form x Concept* showcases product packaging that are made with unusual materials or come in unique shapes illustrative of the product's design concept. A lot of designers not only prioritise preserving the product in the best possible way, but also have environmental concerns that the packaging can bring. The idea is mirrored in the design by using recyclable materials or packages that are entirely self-disposable. For example, Alien and Monkey's *Sand Packaging* (p.110) is only made out of sand by pressing it into a mould. To reveal the gift, just simply break the box apart. Lastly, *Vision x Illustration* examines designs that focus on art object and illustration. Entries in this chapter in particular have successfully achieved an eye-catching result through collaborating with other artists. Intricate drawings, incorporating an artist's signature print or colour are just a few example of what to expect from the section.

Presenting a medley of playful and meticulously crafted special edition designs from around the world, this book features design by people, who appreciate artistic values, care and feel passionate about precise production process as much as the final result. All entries in the book are full of characters, and we wish the readers to pay attention to the message that each designers want their audience to take away from the product. Get inspired, and let the design speak for themselves, proving that there is no rule when it comes to special edition product design.

PLAY

x

IMITATION

PLAY ╳ IMITATION

Play x Imitation features packages that imitate pertinent objects or other designs to create a jocular aesthetic, and products that invite physical interactions. Often, these tactile designs will remind customers of their favourite childhood games, whether it is self-composing a product, personally drawing the label or scratching off the packaging to reveal a hidden design. By adding a playful aspect, designers are instantly building a personal relation with customers from the first sight.

AROMA CUP

nendo converts the sweet pleasures given by enjoying
Haagen-Dazs ice-creams into an aromatic experience.
Exuding the same whiff of vanilla, only by scent, the
studio created an aromatherapy candle as a gift item for
Haagen-Dazs. The ice-cream makers' paper cup, lid and
accompanying plastic spoon are recreated with ceramics in
melting ice-cream looks. When the candle isn't lit,
the spoon becomes an aromatic oil diffuser.

Design: nendo
Photo: Haagen-Dazs Japan
Client: Haagen-Dazs

MARC JACOBS BEAUTY 2014

Aspired to highlight the fun and playful elements of
a new cosmetic collection for Marc Jacobs Beauty,
studio Established reinterpreted the products'
packaging and made an analogy between the act
of beautifying and painting. Lipsticks as well as an
eyeliner and eyeshadow duo were all contained in a
sheeny paint stick-like case that created a new and
fun way to put colours on one's face.

Creative direction: Established
Design: Peter Ash, Nils Siegel
Client: Marc Jacobs, Sephora

SHINAN-JI SEKITEI

With the help of Japanese confectionery artisan Motohiro Inaba, designers Tomonori Saito and Shohei Sawada created an edible zen rock garden, "Ka-re-san-sui", that pacifies the eyes as well as taste buds. Set in a book-sized case are black sesame rocks that can be arranged at will on a bed of sugar, complete with a wooden rack that combs the sweetest ripples.

Design: Tomonori Saito, Shohei Sawada
Confectionery: Wagashi Asobi

LORETTA SWEET COLLECTION

From their pleasant smell to fun character, Loretta's styling products evoke what grown-up women had wish for since they were a little girl. At its fifth birthday, Loretta launched a collection of salon products that emanate a sweet aroma like one would catch at an European pastry store. The dessert-inspired packaging lifted the party spirit with a sense of innocence and playfulness that Loretta has always advocated.

Creative direction: Aya Obayashi (beauty experience Inc.)
Design: Sayaka Koda (Nippon Design Center, Inc.)
Illustration: Hideatsu Morimoto
Client: MoltoBene INC.

CHOCOLATE-PAINT

nendo doubles our childhood excitement of unpacking a
brand new box of paints by replacing the aluminium tubes
with chocolate delicacy. Carried by its product brand "by | n",
different flavoured syrups are filled in as edible paint. Tube
labels are also wrappers that keep kids and kidults' fingers clean.

Design: nendo
Photo: Ayao Yamazaki
Client: by | n

ALT XMAS 2010

New Zealand studio Alt Group moulded 150 full-size keyboards from dark Belgian chocolate for their clients to devour during the Christmas of 2010. An award-winning design, this edible gift is packaged in a plain white case where keyboard indexes are vaguely visible. It is like a word puzzle that leads recipients to draw hints about the gift, except for the Alt key in red that directly refers back to the studio.

Design: Alt Group

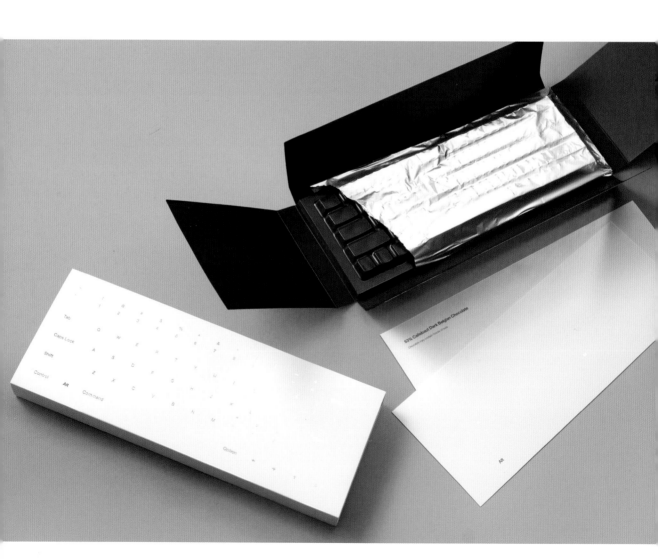

SKIRT-FLIPPING

Skirt-flipping is a humorous desktop calendar inspired by the everlasting schoolboy fantasy of lifting up a girl's skirt. The design illustrates the bottom half of a high school girl wearing a typical school uniform. Have some patience, flipping one skirt-page per month, and a surprise will be waiting at the end of the year.

Design: Kaori Kato
Printing: GRAPH Co., Ltd.
Photo: Takuya Nagamine

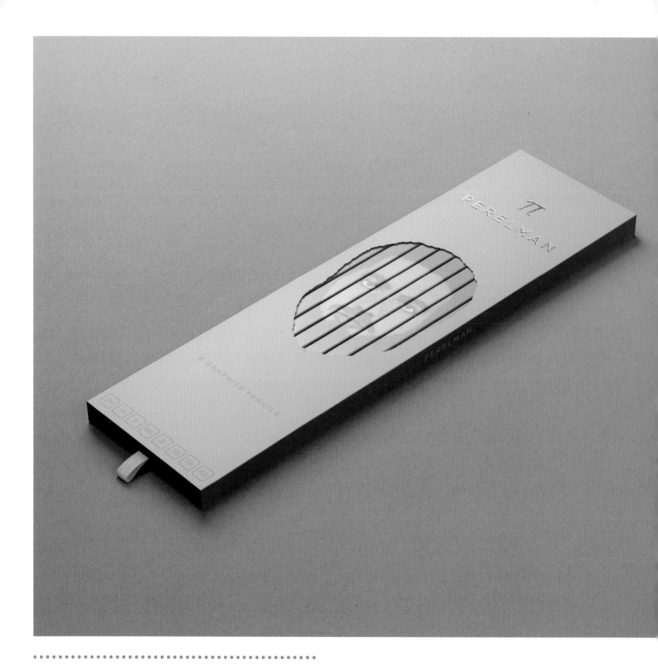

PERELMAN PENCILS

Inspired by the modern Russian mathematician Grigori Perelman, the Perelman Pencils is a studio project by The Bold Studio. Believing money cannot buy the passion they devoted into their profession, which is a notion they share with Perelman, Bold created this set of pencil with Perelman's portrait as a tribute to this honorable genius.

Design: The Bold Studio
Illustration: Jules Julien, Philipp Gorbachev

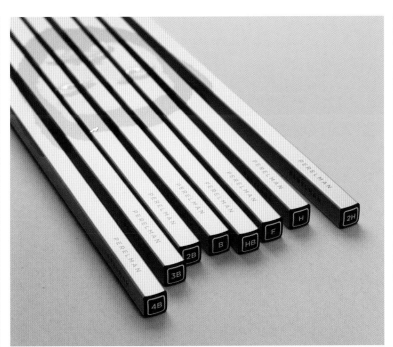

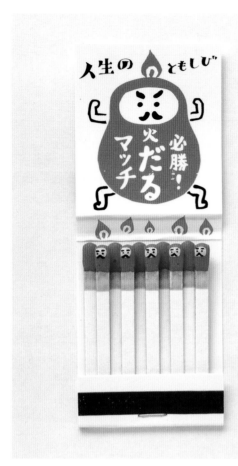

HIYOKO MATCH

Hiyoko Match is originally an idea for a group exhibition in 1994, when graphic designer Kumi Hirasaka used matchbox and matches as materials to create a series of artwork distinguished by tininess. It features traditional Japanese Kokeshi wooden doll on each individual match's head. After several years of hiatus, the matchbox collection is now a souvenir product available in selected stores and online platform.

Design: Kokeshi Match Team

人生のともしび

こけしマッチ

人生のともしび

いぬマッチ

人生のともしび

つるマッチ

人生のともしび

ひよこマッチ

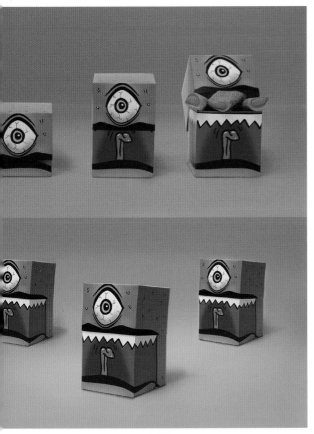

MONSTER CANDY

Initiated by branding and packaging design studio Olsen & Oslo, Monster Candy is intended for the studio's potential clients and friends as a Halloween gift. Featuring three flavours — sour watermelon toffee, banana jelly beans and kiwi flavored sours, each of the boxes depicts a monster with matching colour heads. To create a playful result, the edge of the covers was designed to imitate the mouth of the monsters, as if they were eating its own body parts.

Design: Olsen & Oslo

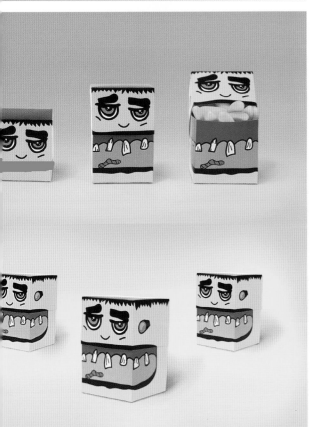

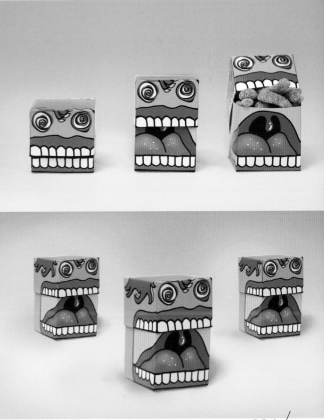

THE LAYER HEN

Designed for Moscow Farmer as a limited business souvenir, The Layer Hen is an add-on handle clad in a colourful illustration of a mother hen. The concept is to demonstrate how fresh Moscow Farmer's eggs are, and how they are handled with care like a hen would with its eggs. Perforated at the bottom, the paper handle can be effortlessly combined with existing egg carton packaging and folded for easy storage.

Design: Andrew Ushakov (Geometry Global)
Client: Moscow Farmer

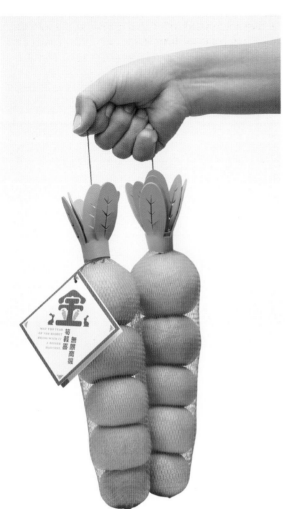

A BIGGER HARVEST

A homonym of "gold" in Chinese, Mandarin oranges often represent "wealth" in Chinese culture. For the year of rabbit, an orange net was used to hold the fruit and fastened by a plastic cap, as a New Year gift sent to wish clients a successful and profitable year ahead. Mandarin oranges were carefully selected by descending sizes to assure the bundle looked like a beautiful carrot.

Design: Alpha245

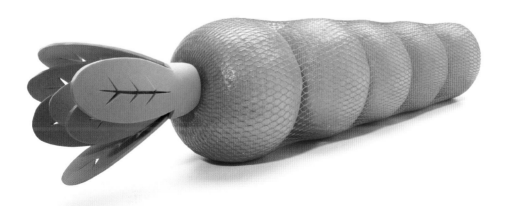

FRUITS TOILET PAPER

Toilet paper rolls have become a novelty gift for Japanese companies and stores to appreciate their customers' support. To turn this "personal" item into a pleasant surprise, Latona Marketing Inc. designed a fresh look to a series of toilet paper by dressing up the paper roll with a realistic fruity wrapper including the skin of kiwi, watermelon and strawberry.

Design: Latona Marketing Inc.

BEER COLORS

Beer Colors is a packaging concept based on the colours of the beer itself. It created a Pantone colour spectrum to differentiate the various flavours, such as Dark Ale, Golden Ale, Imperial Stout, Pale Lager, Pilsner, Porter, Wheat Beer, Shout and Lager. To accompany the colour-driven minimal design, designer Txaber Mentxaka used the Hipstelvetica bold font to add a jazzy personality to the collection.

Design: Txaber Mentxaka
Typography: José Gomes

DARK ALE® 1815 C

PORTER® 1817 C

STOUT® 4975 C

IMPERIAL STOUT® 426 C

DOLINA

Named "Dolina", this craft beer is brewed in Burgos, Spain, where the archaeological site of Atapuerca is located. Drawing reference to the site, the label design wakes drinkers' desire to make discoveries of the beer's delicate taste and the story behind the beer with a tantalising layer of gold ink. Once scratched off, the label reveals a sketch of a human cranium on the front, and a short story about the man's death at the back.

Design: Moruba
Illustration: Marta Zafra
Client: Brebajes del norte

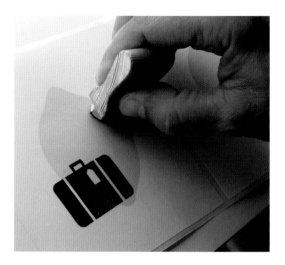

FELICE LIMONE

Literally meaning "happy lemon" in Italian, Felice
Limone produces artisanal limoncello. The idea of crafts
and unique quality is translated into a pared down
label, which the manufacturer can customise the look
with rubber stamps combining graphics and colloquial
expressions. The result is a bunch of happy lemons that
no two of which look the same.

Design: Moruba
Client: Felice Limone

DIN-AMIC

Created for French ephemeral tableware label IPI, Din-amic is a modular tableware collection known for its streamlined finish. Its new cutlery set specially made for Paris' Pompidou Museum Boutique in 2014, is disguised as a model toy plane set that one can actually play with. Coming in an "assemble-yourself" sprue, Eugeni Quitllet injects playfulness into this elegant series.

Design: Eugeni Quitllet
Client: IPI

SHAPE
YOUR
KNOWLEDGE

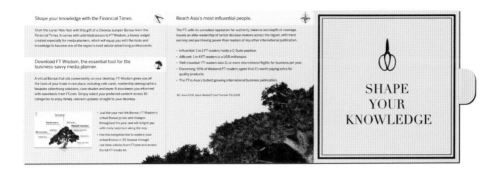

SHAPE YOUR KNOWLEDGE

· ·

Created by Financial Times, FT Wisdom is a handy widget that tailors key information and solution for media planners. Substance built a year-long campaign around a metaphoric Chinese Juniper Bonsai that targeted the tool to Asia Pacific professionals. Genuine bonsai were sent together with a printed manual, illustrating the shaping and growth of both virtual and organic plants, followed by a care package half year later to recapture their attention.

Design: Substance
Client: Financial Times

CHOCOLATE BARCODE

Created as the first of a promotional packaging series, Chocolate Barcode enhances the enjoyment of a quality sweet based on the common word between chocolate bars and barcodes. Only 100 bars were made, and each was numbered on its packet, next to the origin and intensity of the product's cocoa content. The logo also reflects the chocolate design with graphics molten into stripes and lines.

Creative & art direction: Denis Lelić
Project management: Ajla Hamulić

BABAU

Designed for a Catalan wine trilogy hosted by wine distributor Wine Side Story, Babau 2012 was a white wine product produced by cellar Babau. Carrying no specific name, the wine label features a minimal die-cut water drop graphic that invites drinkers to interpret its meaning as they taste the wine.

Design: Lo Siento
Client: CUVEE 3000/ JOAN VALENCIA

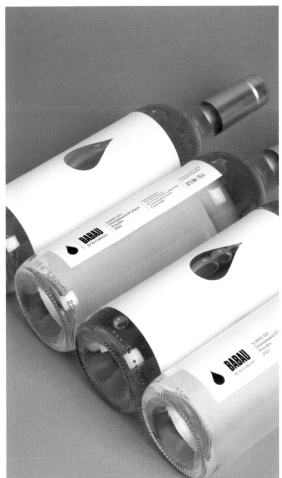

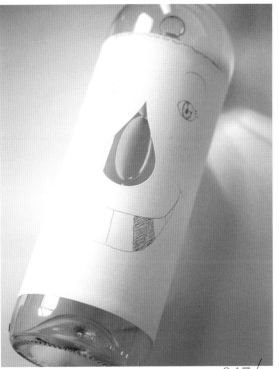

SPINE VODKA

This minimally designed high quality vodka bottle represents the product identity through its entire design. The transparency suggests a product that does not have anything to hide and the gold spine in the bottle describes the liquor a product with a "backbone", promising an honest and proud work to put in front of customers. On the side of the bottle is an attachment of a short and zesty description of the vodka, with the serving instruction.

Design: Johannes Schulz

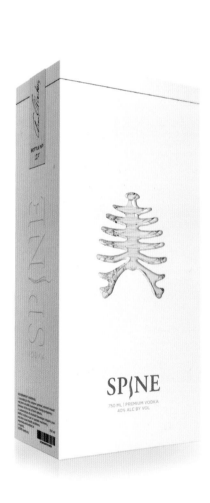

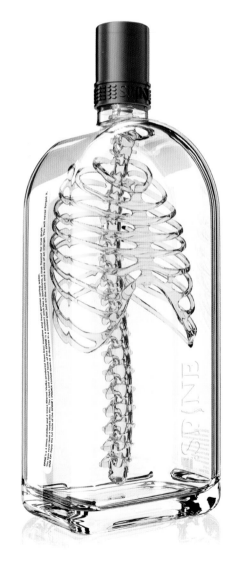

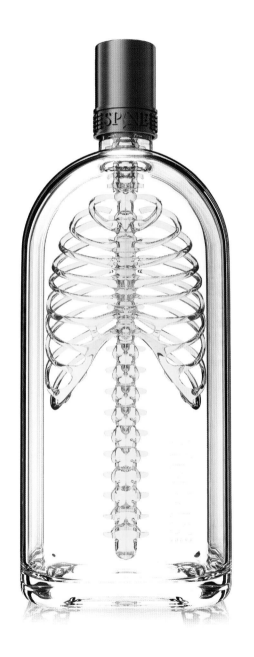
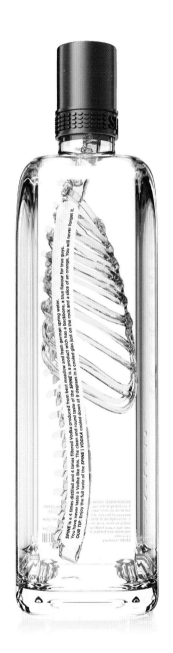

ÉCORCHÉ

écorché is a concept project for an energy drink brand. Inspired by the French Renaissance architect and theorist Leon Battista Alberti, designer Constantin Bolimond found an explicit approach to present the abstract feelings of energy and power. By illustrating the meticulous flesh on the bottle, it reflects the spirit of Aberti's suggestion that the best way of drawing a lively object is how the painter captures the details of the inside.

Design: Constantin Bolimond

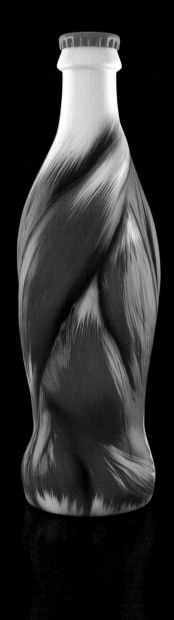
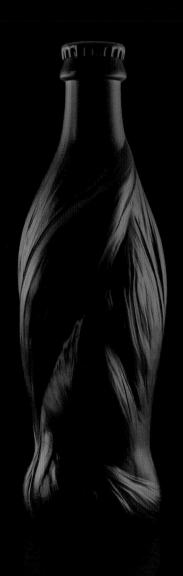

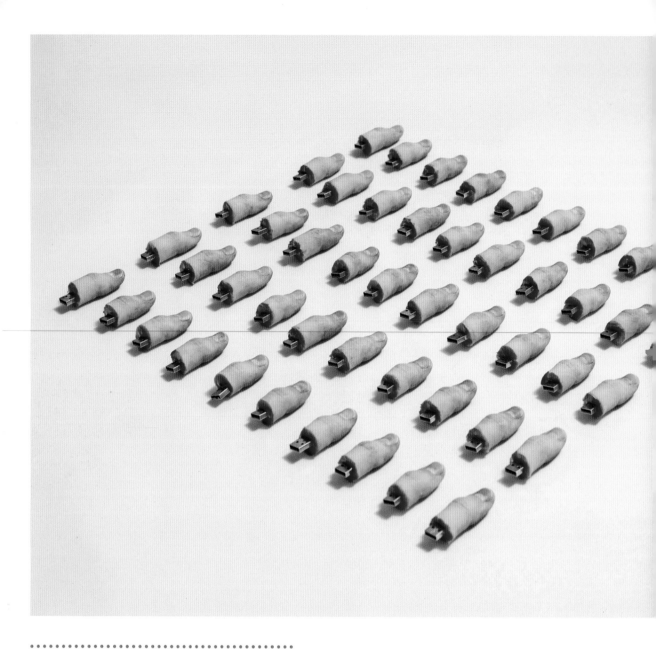

THUMBDRIVE

Justin Poulsen is a conceptual photographer, who wants to shy away from abusing digital wizardry. To present his portfolio and make an impression on his potential clients, he came up with "thumb" drives project. The USB is a hyper realistic sculpture of a broken human thumb that lets the viewers experience both the tactile and visual quality of his work.

Creative direction: Ian Grais, Chris Staples (Rethink)
Art direction: Hans Thiessen (Rethink)
Photo & client: Justin Poulsen

Stick out like a sore thumb.

DRAGON TANGRAM

To celebrate the arrival of the year of dragon, this Chinese New Year gift box set is ideal for family gatherings. Taking tangrams for reference, the design carries seven geometric units that can be configured into Chinese auspicious animals and reveals a dragon when all is fitted into a square box, with a Chinese word "Spring". A quiz card is included to suggest possible combinations with the pieces.

Design: Yin Tsan-yu
Client: Xue Xue Institute
Special credits: Rebecca Hwang, Kuo H-Yun

OBLIQUE

DEREK WELSH STUDIO
GRAPHICAL HOUSE

Paul Smith

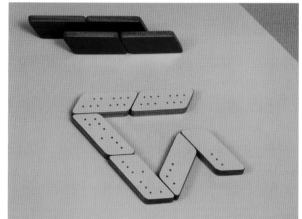

PAUL SMITH OBLIQUE

More than just a spruce edition of the classic domino game, Paul Smith Oblique is an absolute joy to touch and play with. With an edition of 50 sets, Paul Smith Oblique is a collaboration between design consultancy Graphical House, furniture maker Derek Welsh and fashion brand Paul Smith, embodied in the union of fine craft, remarkable packaging and Paul Smith's signature palette. The "oblique" theme runs seamlessly through the walnut bones with hand-drilled spots, all held in an oblique box and drawstring bag.

Design: Graphical House, Derek Welsh Studio, Paul Smith
Client: Paul Smith

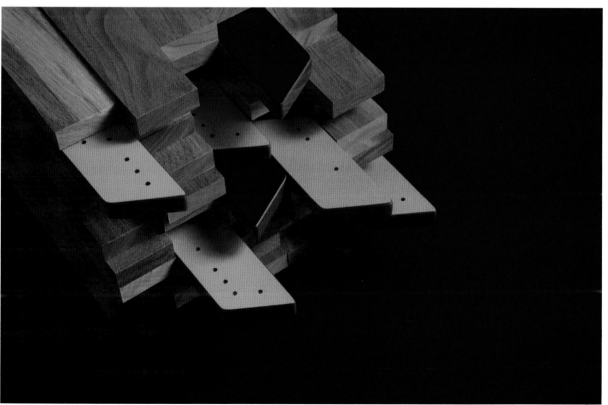

PEOPLE BLOCKS

Jean, Antoine, Fabienne, Francois are four wooden
sculptures, all made and painted by hand. The artist Andy
Rementor's spirit is accurately captured on these rather
pensive characters, by using distinctive patterns and bold
colours. Only 12 were made for each character, the most
important of all is that customers can create an abstract
sculpture by personally stacking them. The pieces are
interchangeable and dependent on the arrangement, the
characters give out a playful and humorous expression.

Design: Case Studyo x Andy Rementer

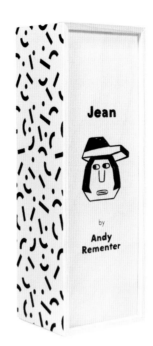

Jean

by
Andy Rementer

PEOPLE
BLOCKS

by
Andy Rementer

VIBES MELT DOWN 2043

A collaboration between graphic artist Cody Hudson and Case Studyo, Vibes Melt Down 2043 is an art piece as well as an incense burner. Sculpted into the shape of a cranium, the porcelain sculpture aims to evoke enlightening moments of life and death when an incense burns inside. Each ware comes in a wood box with a signed and numbered certificate by the artist and three cones of incense.

Design: Case Studyo x Cody Hudson

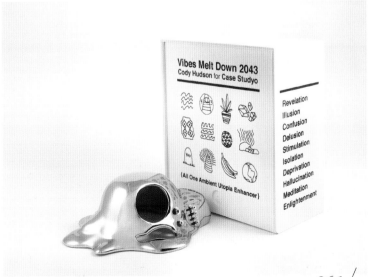

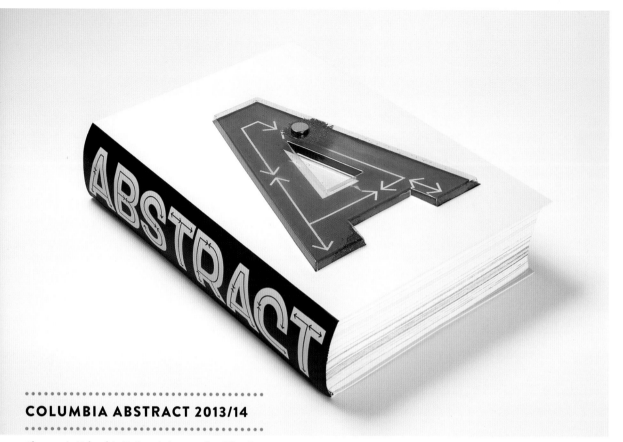

COLUMBIA ABSTRACT 2013/14

Abstract is Columbia University's annual publication for the Graduate School of Architecture, Planning, and Preservation. Featuring an A-shaped PVC case that encapsulates iron shaving, which can be moved around by using the accompanying magnet, the 2013/14 edition's cover metaphorically describes how the school "attracts" students to study in the New York City before "repelling" them into the world to experience architecture worldwide.

Art direction & design: Sagmeister & Walsh
Client: Columbia University

EVIAN STAR WARS

The Star Wars edition is a conceptual collaboration between French mineral water brand Evian and the epic sci-fi masterpiece Star Wars. More than just featuring the film characters like Darth Vader, Stormtrooper and the astromech droids in geometric style, the glass bottle was designed as a hilt of the iconic laser sword Lightsaber.

Design: Diego Fonseca
Client: Evian

2013 NEW YEAR LIMITED EDITION WINE BOTTLES

To celebrate the new year's arrival, ThoughtAssembly created limited wine bottles with 'animated' sleeves, achieved by overprinting two striped numerals on the label and a transparent outer sleeve creating an illusion of "12" that changes into "13". Produced and given away to demonstrate the studio's competence to communicate in visuals, the bottle was paired with letterpressed coasters, a still interpretation of the idea of change.

Design: ThoughtAssembly

POINT AFTER POINT WINE BOTTLES

At an Annual PetFood Forum Tradeshow, pet food manufacturer Cargill intrigued its potential customers with a connect-the-dots game. When joined using a white colour pencil that comes with a red wine bottle, the dots on the wine label reveal a feline or a canine, which explains Empyreal 75, a star product which Cargill has developed for pets. Once that subject is brought to light, there's reason to celebrate — with a nice glass of wine.

Design: Bailey Lauerman
Illustration: Greg Paprock
Client: Cargill

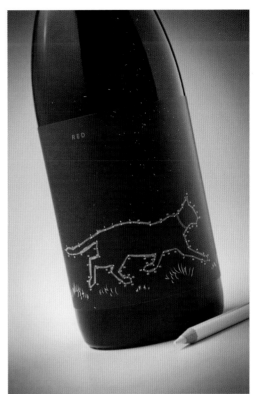

NEWS OF THE WOOLED

Introduced as a knitting kit, News of the Wooled emphasises the functionality of packaging design. Coming in pairs, and with an opening at the centre, the packaging cards can stand like a sheep as they hold a woollen yarn ball — a graphic reference to the fibre's origin. Also, the set offers knitting needles with tip covers and yarn bobbins.

Design: Gwyn M. Lewis

COTTON TWITTER

Sealed in the shiny tin is a cotton shirt with a writable surface and three pieces of chalk sticks where one can use to scribe their mood or a message for display. The special paint layer will allow these messages to stay as long as the one want, and be cleared out as it's washed with water.

Design: MARCH design studio

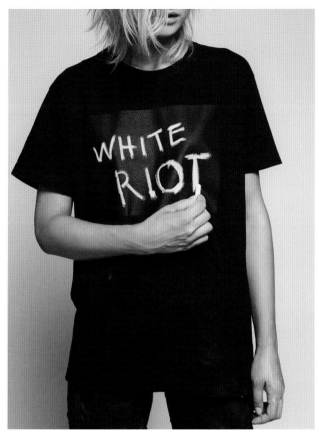

ALL FIT

To manifest adidas' forte of catering sportswear and accessories perfectly to the needs of their core athletes, the studio fitted the season's tailored products for each of them into bespoke cases. Items in the Spring/Summer 2013 All Fit box were embedded individually for clear display, covered by a Lucite board that illustrates personalised performance metrics and how every product is custom-fitted to the athletes' body.

Design: Caviar Digital
Client: adidas global training division
Special credits: Hans Cheung (adidas training division)

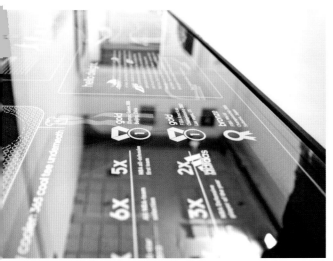

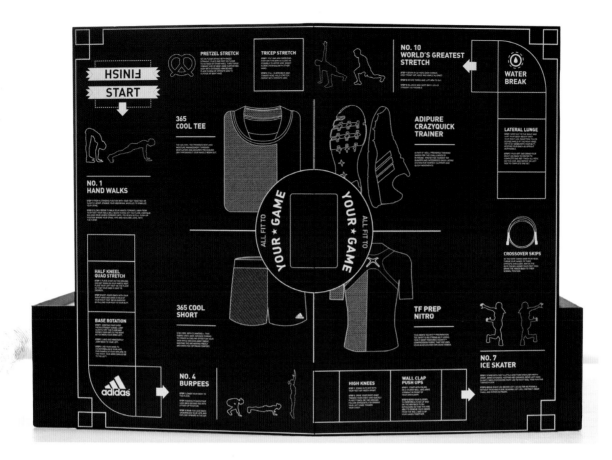

ALL FIT TO YOUR GAME

Evolving from last season's customised showcase, the
studio turned each of adidas core athletes' Fall/Winter
2013 package into a tailored board game set. As a token
for the recipient to be on top of their game, training
instructions and tailored gear are illustrated like game
steps in the two-piece box. Starring as joker in the
enclosed deck of cards, the athlete might want to roll a
dice or two for a more exciting game.

Design: Caviar Digital
Client: adidas global training division
Special credits: Hans Cheung (adidas training division)

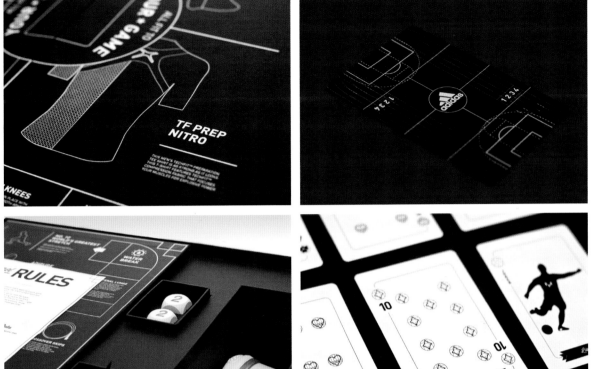

UNDER ARMOUR®
SPEEDFORM APOLLO SEEDING BOX

'Speed' is visualised as a jet that figuratively describes how Under Armour's SpeedForm Apollo running shoes are optimised to enhance a runner's speed. The custom packaging was specially designed for the "THIS IS WHAT FAST FEELS LIKE" shoes launch campaign. With a transparent window, collectors will see the shoes neatly sitting in the pilot seat alongside a jet-shaped USB keychain, a card and a tee.

Design: Chi-Chi Bello
Client: Under Armour
Special credits: SMS Display Group

ROBOT ROY NUTCRACKER TOY

For Christmas 2014, Robot Food gave their clients, family and friends a festive wooden nutcracker with a robotic twist. Robot Roy was hand-painted in primary colours and stored in a bespoke black box made from Ebony Colorplan, with a rendered gold-foil robot on it. To boost a sense of exclusiveness, a personal note was inserted into each of the 50 limited edition box, sealed with a numbered sticker.

Design: Robot Food
Special credits: G.F. Smith

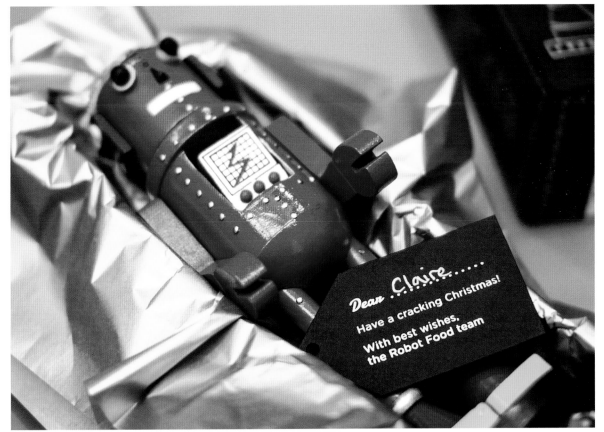

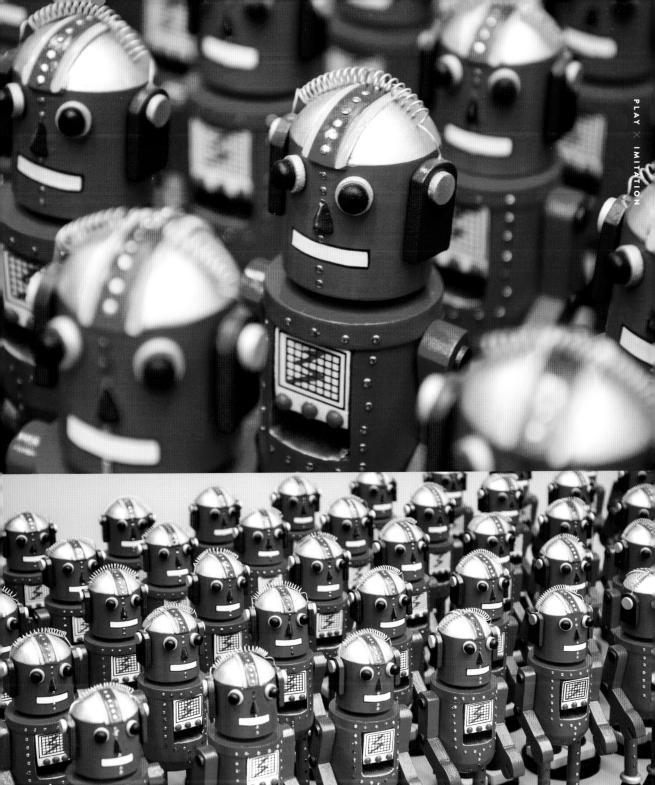

FORM X CONCEPT

FORM $\times$ CONCEPT

Form x Concept presents product packages that are made with unusual materials or come in an unconventional form. The materials can be anything from rock, ice, feather and glass; often mirroring materials into the production or the equipment involved. Escaping the traditional square box packaging, these packages often reflect functionality of the product, or even the brand identity. With no restrictions in how they should be presented, the styles they come in are as diverse as the designer can stretch their imagination.

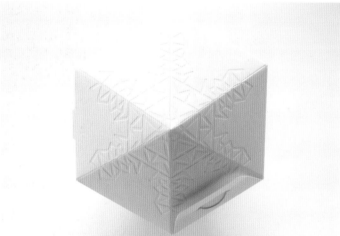

TOWADA PREMIUM GARLIC

Towada Premium Garlic is a local specialty from Towada city, packaged as a souvenir product for the Towada Art Center. To make it a memorable gift item, designer Keiko Akatsuka created a snow crystal-shaped box that reflects the snowy mountains in the Towada City as well as the pure white art hub. To keep the whole package as pure as snow, she decided to deboss the information and place a snowflake graphic on the box.

Design: Keiko Akatsuka & Associates
Client: Oiseau Inc.
Special credits: Towada City, Aomori

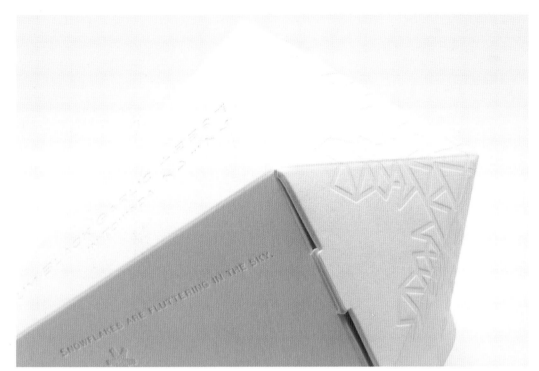

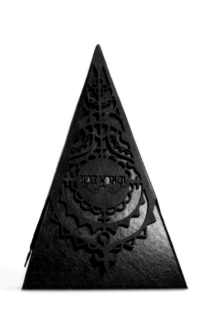

KIDROBOT KAT VON D EDITION

KidRobot created a special edition of toy figure packages inspired by the famous tattoo artist Kat Von D. Carrying the same gothic-elegant style that embeds in her tattoo, makeup and fashion line, the toy figure packaging uses black on black, with an intricate pattern achieved by laser cut.

Design: Wang Mei-cheng

EXTREM DELUXE

With a long history of producing premium ham, Spanish company Agriculturas launched a new Iberian, acorn-fed ham brand, EXTREM. To match the product's quality and the brand's desire to reposition itself in the high-end food market, a prestigious matte black container box was conceived to hold the product and function as an elegant serving plate. A golden piggy handle highlights the content and makes the product a natural gift choice.

Design: Lavernia & Cienfuegos Design
Client: Gallén-Ibáñez, AGR! for Agriculturas Diversas SLU

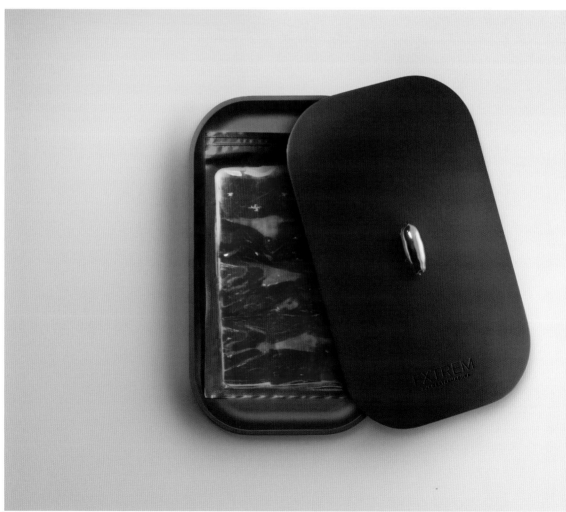

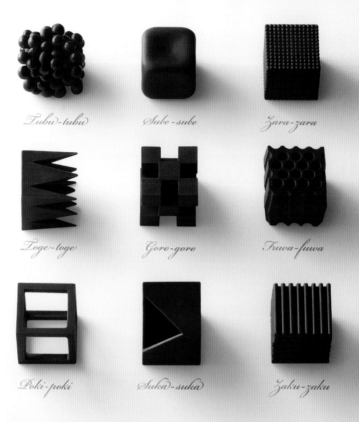

Tubu-tubu

Sube-sube

Zara-zara

Toge-toge

Goro-goro

Fuwa-fuwa

Poki-poki

Suka-suka

Zaku-zaku

CHOCOLATEXTURE

Crafted for lifestyle and interior trade fair Maison et Objet in 2015, nendo's chocolatexture is an experiment about how form and texture changes the sense of taste. Made from identical ingredients, each of the nine luscious sculptures introduces distinctive personalities to the taste buds within a 26 mm cube dimension. Limited sets were available for purchase and to taste at the on-site chocolatexture lounge, where nendo's previous works were custom-coloured to resemble molten chocolate.

Design: nendo
Photo: (chocolatexture) Akihiro Yoshida,
*   (chocolatexture lounge) Joakim Blockstrom*
Client: Maison et Objet

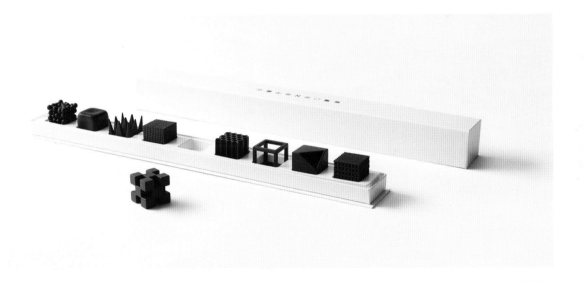

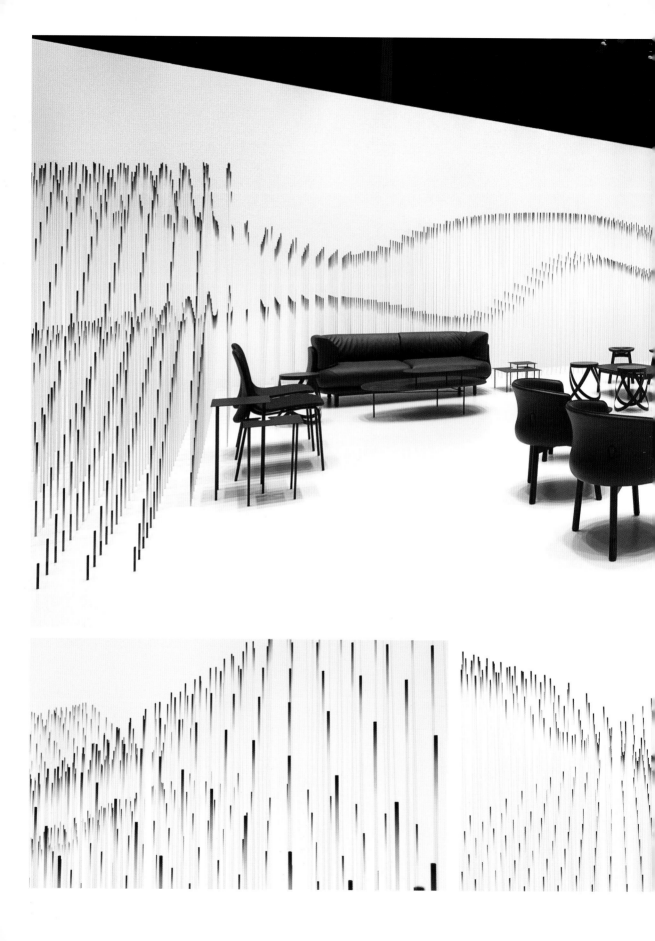

TAKESHI MURATA SLIPCASES

Unique edition of slipcases were created for three of Takeshi Murata's videos namely Shiboogi, Night Moves and Street Trash. Each case includes three USB drives of Takeshi's video works embedded in die-cut archival foam and a certificate of authenticity. The slipcases were fabricated via selective laser sintering in nylon plastic.

Design: Familiar Studio
Client: Takeshi Murata, Salon 94

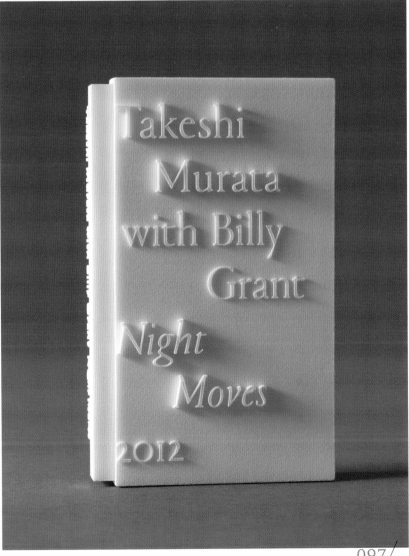

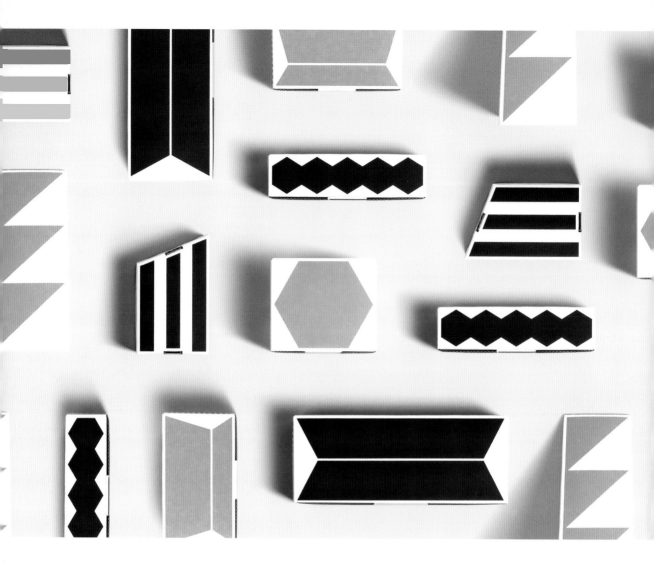

BASALT

Supported by the Korean Craft and Design Foundation, Basalt is a six piece set of geometric stationery line and office supplies made out of stones carved in Jeju. From the hexagonal coasters that mimic the Korean island's coastal rock formations, to the triangular bookend that reflects the 386 volcanic peaks there, each is a natural stone creation sourced and crafted in Jeju and individually packaged to emphasise their geometric nature.

Design: Seo Jeonghwa

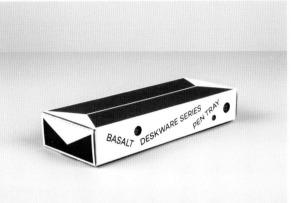

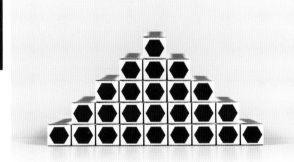

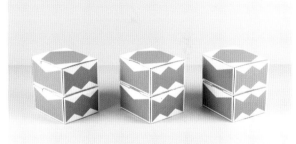

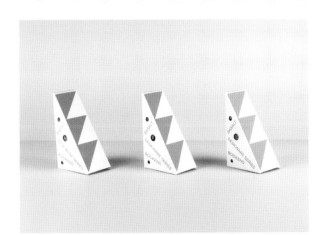

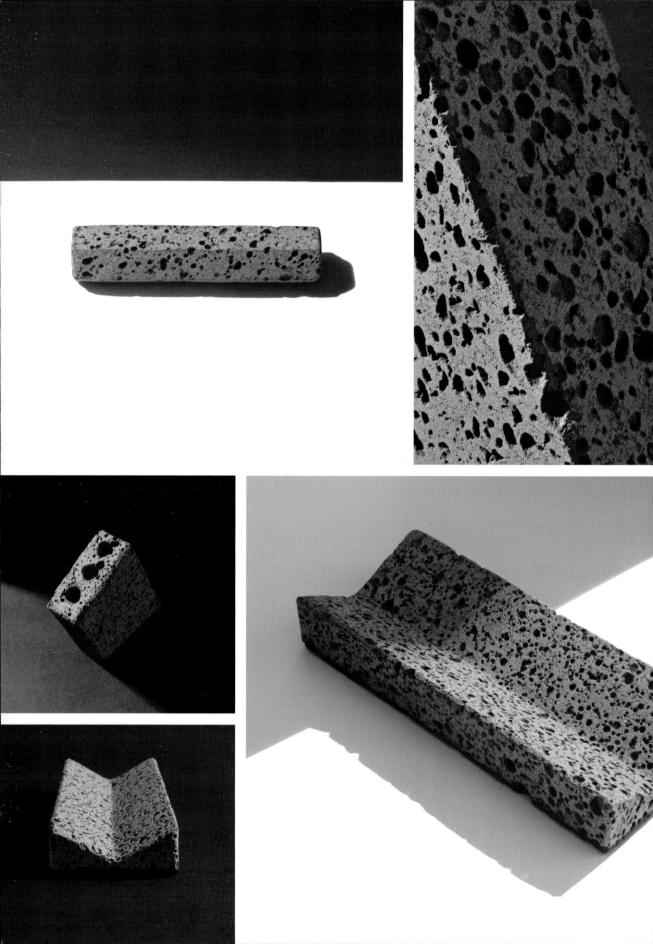

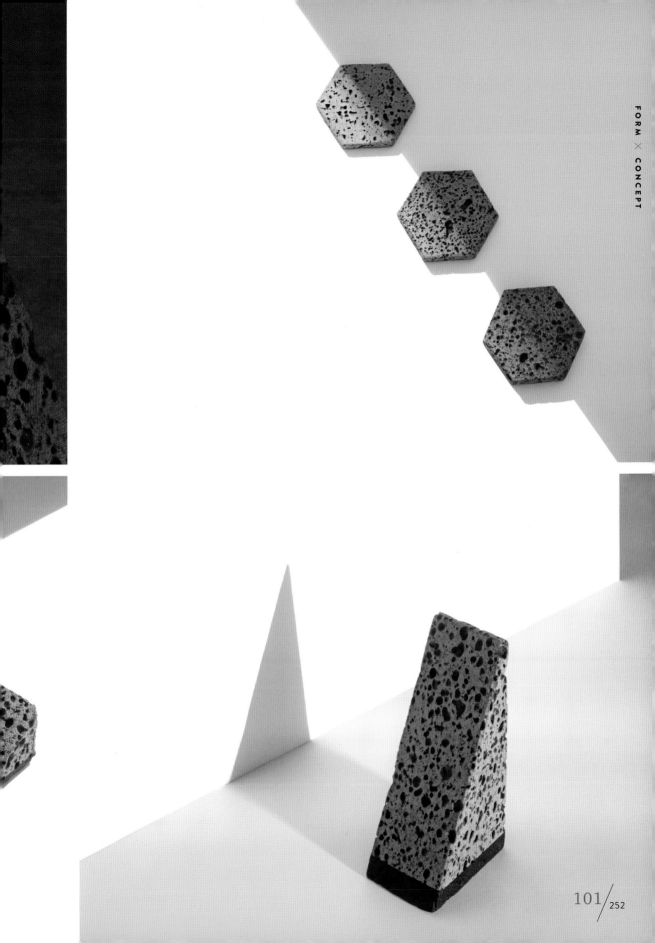

Eau de Parfum
60ml 1.6 FL. OZ.

BARB PERFUME

Barb men's special edition perfume bottle is not typical in any way. The bottle itself is a stone and the flask with perfume is placed inside it. There is no obvious tags or logos attached, because the overall philosophy behind the design is to represent freedom, fearless spirit, strength and simplicity. Through the design the Barb team is expressing there are no boundaries and our spirit is strong enough to break through concrete walls.

Design: Barb Team

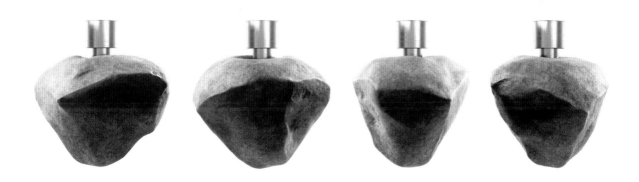

RELEASE YOUR NATURE

WILDBARB.COM

RELEASE Y

WIL

RELEASE YOUR NATURE

WILDBARB.COM

THE TEA CALENDAR

Comprised of 365 calendar day plates made from tightly
pressed tea leaves, Tea Calendar invites Hälssen & Lyon's
business partners to enjoy everyday with a cup of Hälssen &
Lyon tea. Once a day, they are expected to pluck a sheet from
the calendar box and steep it into hot water for an exclusive
taste offered by Hälssen & Lyon's creative blends. Each
calendar sheet were hand-cut and stamped with days and
months, which are appropriate for consumption.

Design: Kolle Rebbe GmbH
Client: Hälssen & Lyon

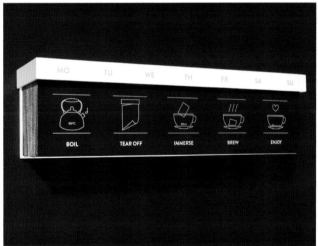

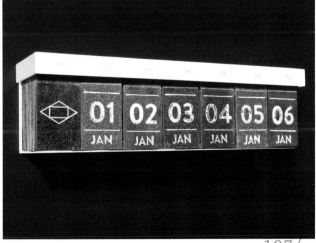

MENZO SOAP

Menzo Soap is a packaging design intended for Menzo's male customers. In every detail, it is meant to accentuate masculinity. Made out of concrete, the overall shape is bold and sharp. The wooden cover is made out of Taiwanese cypress, subtly underlining the brand's root.

Design: Lin Yu-heng

SAND PACKAGING

alien and monkey's sand packaging creation takes a very different approach on unwrapping a gift box, by using one of the most abundant raw materials in the world, sand. The box is moulded with the granular substance. Small items such as rings and pens can be stored. It re-establishes the feeling of discovering something, reminds people of much simpler and innocent times. There is also no concerns for waste level, since it is made with all natural resources.

Design: alien and monkey

SO DAMN PROPER
LIMITED EDITION COASTERS

Furnishing company So Damn Proper's 2014 product line went back to the basics. It was an attempt to capture the finer things in life through clean modern design. To expand the collection, VAUX helped the company to develop and package six sets of wood coasters that echo natural earthy tones of the season. Each limited edition coaster was laser-etched with motivational wisdom that will make one's day.

Design: VAUX

OAK WINE

OAK wine is a collaboration project between design studio Grantipo and ad agency La Despensa. In order to prevent unwanted further fermentation of the wine after the production, the team created a bottle made of the same wood that was also used during the process of wine making. From beginning to the end, the packaging helps preserve consistent flavour, because the wine is exposed to the same habitat.

Design: Grantipo, La Despensa
Client: OAK wine

THE REAL ICE COLD COCA-COLA

Coca-Cola has been consistently satisfying public's thirst with their refreshing beverage since 1886. To elevate their game and the enjoyment of coke, Ogilvy & Mather Bogotá created limited edition coke bottles that were entirely made out of ice, letting the customers enjoy cold and refreshing coke until the last drop. An added advantage is that the ice bottle is beach friendly, great for the environment by leaving no waste behind since the packaging disappears after the disposal.

Design: Ogilvy & Mather Bogotá
Photo: Dario Mora
Client: Coca-Cola

A LEAN YEAR

Multidisciplinary design studio Alt Group created a
special edition gift for their client. It was a reward for
themselves for getting through the year 2009, which
was one of the worst years in global economic history.
Detailed thoughts went into creating this cleanskin
bottle of wine. It was slumped, filled and corked
individually by hand, both the content and exterior were
personalised.

Design: Alt Group

A LEAN YEAR

2009

THE HAPPY SHOW
LIMITED EDITION PACKAGING

Limited edition packaging were created for three typographic films, which were part of the travelling exhibition called "The Happy Show". Only ten boxes were made for each film, and every single box was handwritten by creative director Stefan Sagmeister. The box encases a blu-ray disc and a certificate of authenticity, and contains an earthenware USB drive tailored to each film.

Design: Sagmeister & Walsh
Client: The Happy Show

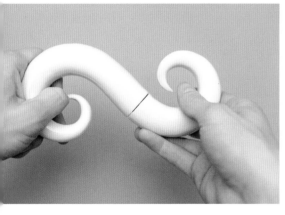

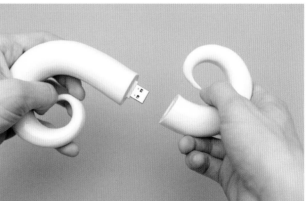

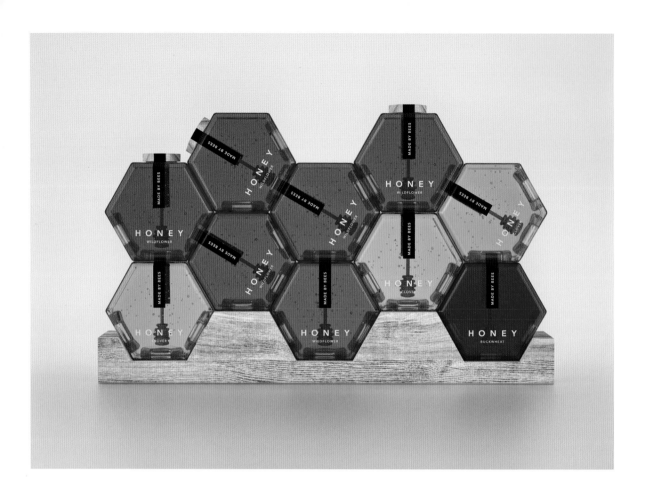

HONEY CONCEPT

Simplicity is definitely not underrated in this project. To deliver the maximum smell and taste of honey, the product is packaged in the simplest and cleanest design, being honey in its own form. The designer wanted his audience to look at honey as nothing more than what it really is, and appreciate it in its natural form.

Art direction & design: Maksim Arbuzov
3D visualisation: Pavel Gubin

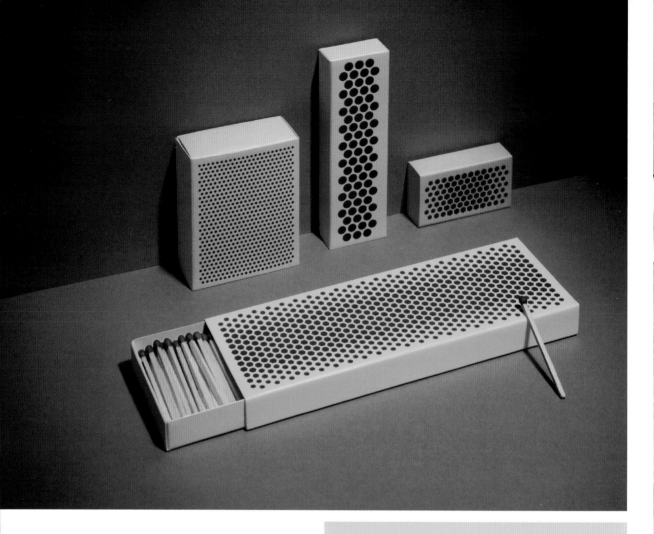

STRIKE MATCHBOXES

Strike is a tribute to safety matches with a red phosphorus striking surface, a Swedish invention dating back to 1844. The inflammable substance appears in geometric patterns, replacing the normal adverts and becomes the prominent feature of each vibrant modular block of matchbox. While most matchboxes are simply put away, Strike is designed to be notably present in the domestic landscape.

Design: Shane Schneck, Clara von Zweigbergk
Client: HAY DENMARK

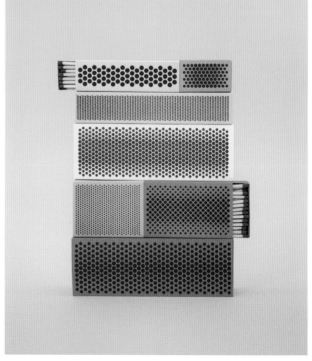

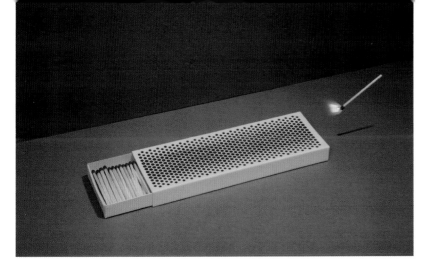

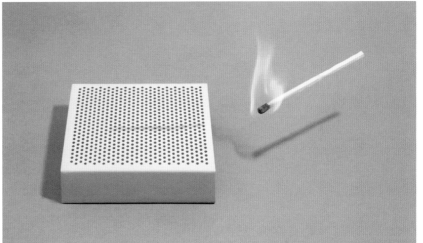

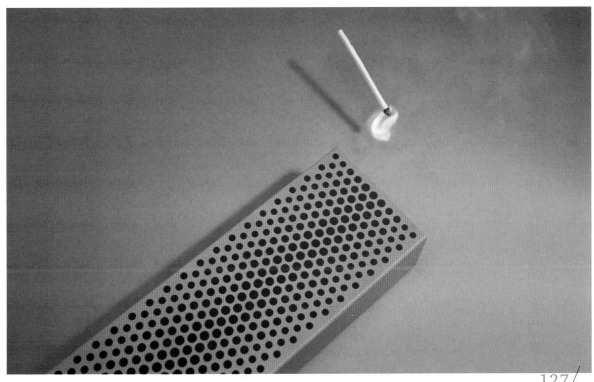

BASIK

Grooming product market is highly customised to gender stereotype. Originated from designer Saana Hellsten's thesis on gendered visual language, Basik is a statement against such inequality in the industry. The sole focus of the gender neutral packaging is the functionality of the product, rather than gender stereotyping. As a result, customers get to choose their product based on its purpose.

Design: Saana Hellsten
3D modelling: Eugene Kim
Client: Pratt Institute
Special Credits: Warren Bernard

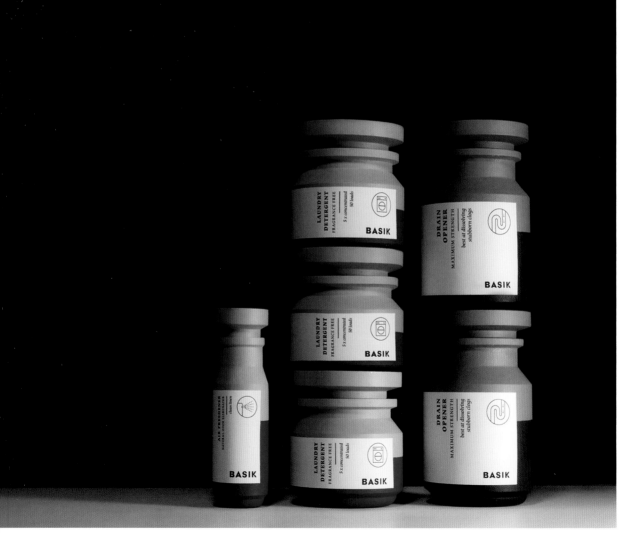

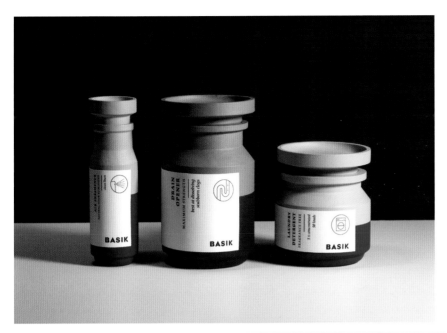

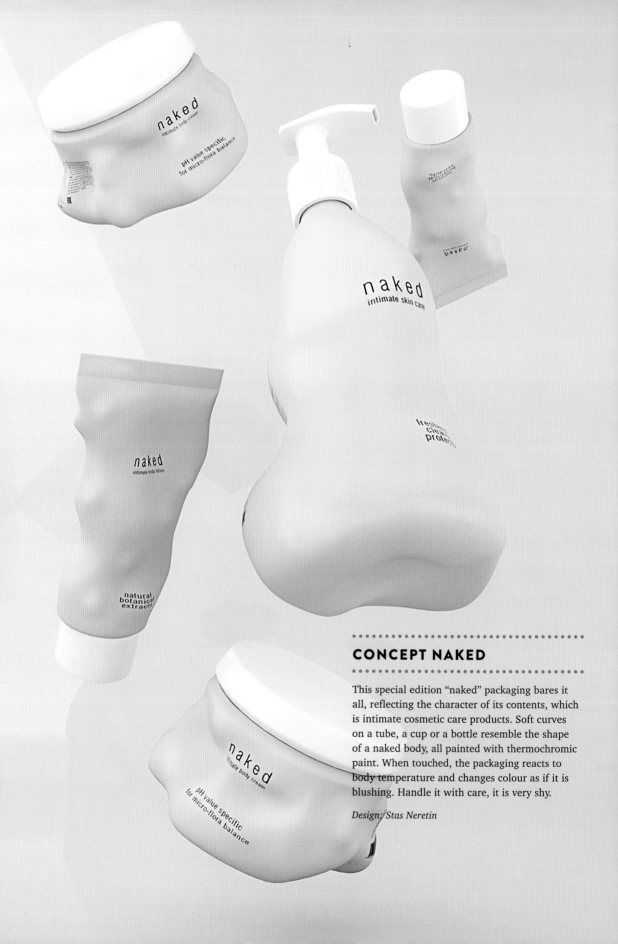

CONCEPT NAKED

This special edition "naked" packaging bares it all, reflecting the character of its contents, which is intimate cosmetic care products. Soft curves on a tube, a cup or a bottle resemble the shape of a naked body, all painted with thermochromic paint. When touched, the packaging reacts to body temperature and changes colour as if it is blushing. Handle it with care, it is very shy.

Design: Stas Neretin

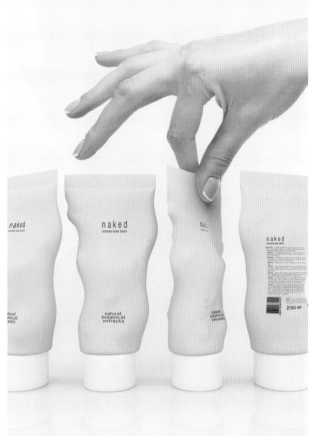

PILLIAD ECHONS

Inspired by the band title, customised CD cases for Greek band Wall to Wall Carpeting were made from scratch by cork, a basic insulating material. The six different cover designs combine into a full-scale single image of an underground river. It underlines the theme of fragmented continuity in the album, which builds a series of melodic structures evolving around one continuous musical form.

Design: Bend
Client: Orila Records

THE GROTESQUE IS BEAUTIFUL

Created as a student project, designer Shar Biggers'
Alexander McQueen home collection packaging is heavily
reflected on the fashion designer's view on life and
definition of beauty. McQueen often valued what the society
thought was grotesque and drew inspiration from it. Pigeon
feathers were the main focus in this design. Dyed black and
stuffed on the exterior, they came from a creature that is
often repelled, but intelligent and strong in reality.

Design: Shar Biggers

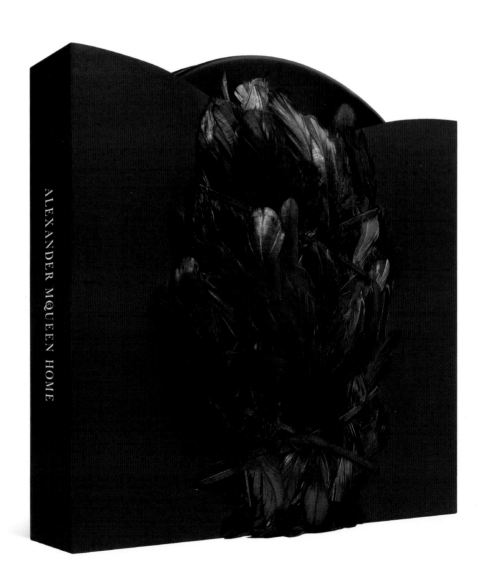

ALEXANDER MQUEEN HOME

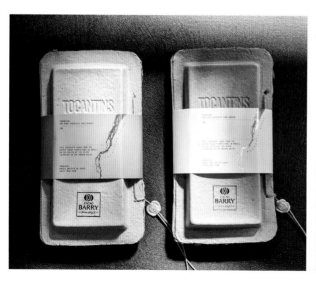

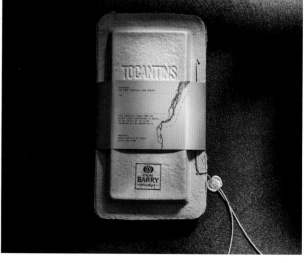

TOCANTINS

Tocantins is a chocolate that comes from small cocoa plantations located in Amazon Delta and produced by chocolatier Cacao Berry. As an official partner of the World's 50 Best Restaurants Awards, Cacao Berry commissioned Zoo Studio to design a special Tocantins packaging for the event. The exclusive packaging was composed with paper paste, individually hand tied with fine rope and sealed with sealing wax, a reminder of Tocantins' remote origin.

Design: Zoo Studio
Client: Cacao Berry

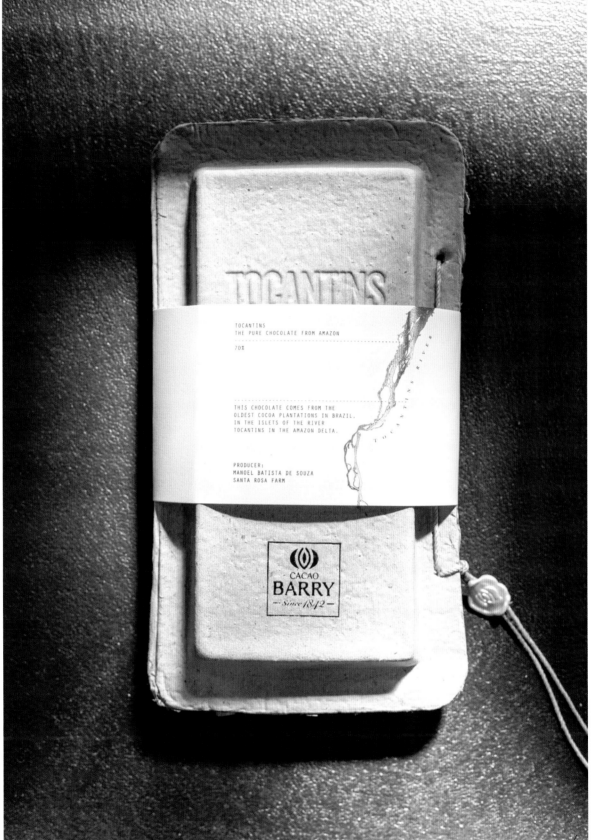

TOCANTINS
THE PURE CHOCOLATE FROM AMAZON

70%

THIS CHOCOLATE COMES FROM THE
OLDEST COCOA PLANTATIONS IN BRAZIL,
IN THE ISLETS OF THE RIVER
TOCANTINS IN THE AMAZON DELTA.

PRODUCER:
MANOEL BATISTA DE SOUZA
SANTA ROSA FARM

CACAO
BARRY
—Since 1842—

ONE PERCENT

Packaging for One Percent sneakers was created with an intention to craft a reusable box that is approachable for their sophisticated customers, the 1%. The long boxes were composed with wood and thick recyclable cardboard. They can be pieced together at an angle and a thick acrylic band in both orange and blue hugs in the middle where these two materials meet. The logo and title are intentionally separated. The rest of the design is simple and clear, to give it a symmetrical balance.

Design: Ryan Romanes
Photo: Vinesh Kumar

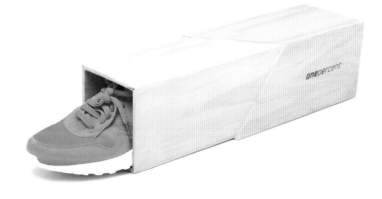

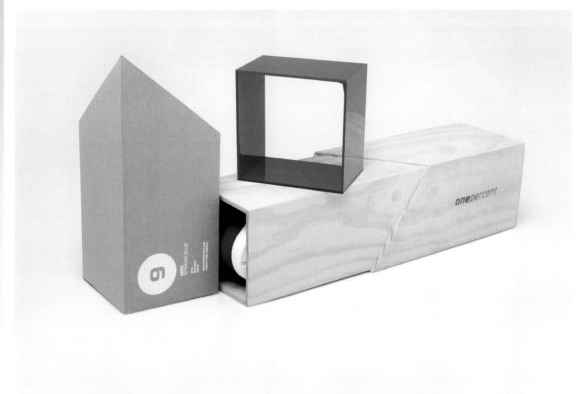

THE WONDERFUL WORLD OF
ALBERT KAHN

The Wonderful World of Albert Kahn is a BBC documentary photography series showcasing Kahn, a French philanthropist sending photographers to document our planet life from 1909 to 1931. It consists of huge volume of letters, diaries and most importantly, film tapes. Specifically for the Polish edition of the collection, Studio Otwarte designed a telescopic box that referenced the shape of early bellow cameras.

Design: Studio Otwarte
Client: M2 films

 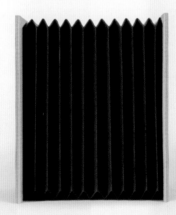 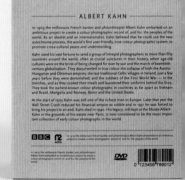

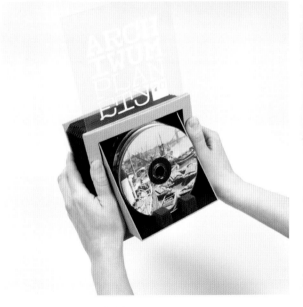

STRELKA TOOLBOOK

STRELKA ToolBook comes as a smart solution to utilise space. Slim and handy, this literal combination of "tools" and "book" is fit for storage in a bookshelf and has a capacity for four screwdrivers and six hex wrenches. STRELKA issued a special white edition to commemorate the launch of ToolBook as their first product. Blind debossed features outside the case not only acknowledged users about its content, but also the story behind STRELKA's brand name and slogan.

Design: STRELKA
Photo & marketing: b.Logic Branding Consultancy Co.

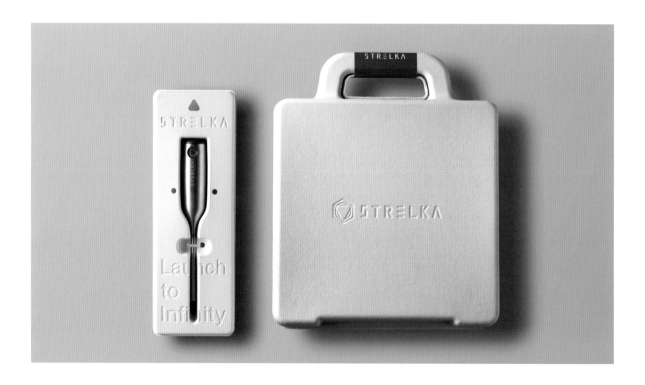

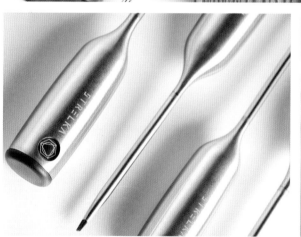

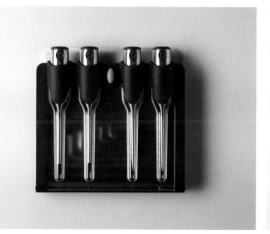

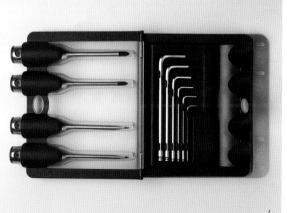

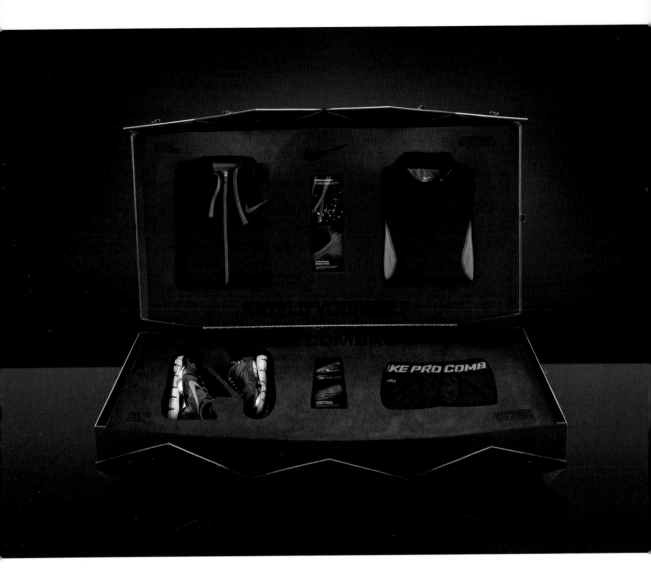

NIKE COLD WEATHER CASE

To connect their NFL pro athletes with wider range
of audience and light up social media, Nike Football
commissioned WSDIA to create a special edition case to
deliver Nike's latest winter products to the athletes. Made
of Dibond/Alucobond, the exterior is a mix of triangular
geometry and battle armor in duffle bag form. On the inside,
CNC waterjet cut foam allowed perfect fit of every product.

Creative direction: Kevin Wolahan (Nike)
Design: WSDIA | WeShouldDoItAll
Fabrication: Brooklyn Guild
Photo: Mo Daoud
Post-production imaging: Portus Imaging
Client: Nike Football

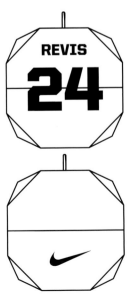

MAŁOPOLSKA SELECTION

The Marshal's Office of the Małopolska Region initiated a project to develop a set of souvenirs composed of over a dozen objects created by local craftsmen. Each object is unique and captures the authenticity and characteristic of the Małopolska region. The packaging itself was inspired by the "Cracow Workshop", a community that was formed in 1913 and devoted to improve the artistic quality in craftsmanship.

Design: Studio Otwarte
Client: The Marshal's Office of the Małopolska Region

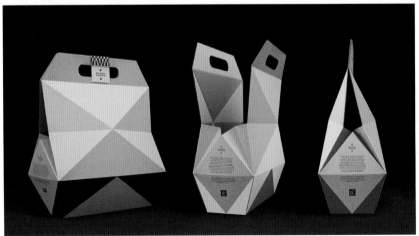

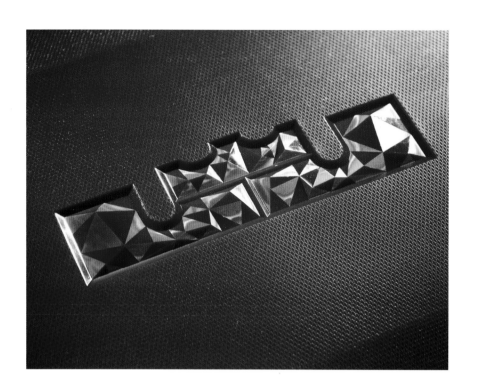

LEBRON

To celebrate the tenth year since launching of LeBron James' signature shoe line, Nike created a limited edition shoebox. A subtle reference to the championship ring, the shiny, diamond-shaped box is intended to pay tribute to LeBron's awards and merits over the years. Geometric facets composed a fascinating structure, while maximising the functionality and practicality of the box.

Design: Travis Barteaux, Nike Inc.
Photo: Ryan Unruh Studio
Client: Nike

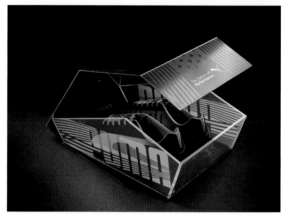

PUMA EVOSPEED

A special packaging was created for the launching of PUMA's new evoSPEED football boots. The shape of shoebox is emulating a stealth bomber, which accents the speciality of the shoes, the ability to maximise speed. Etched by laser, this fold back box is self hinged and has a glowing acrylic edge.

Design: Everyone Associates
Client: PUMA

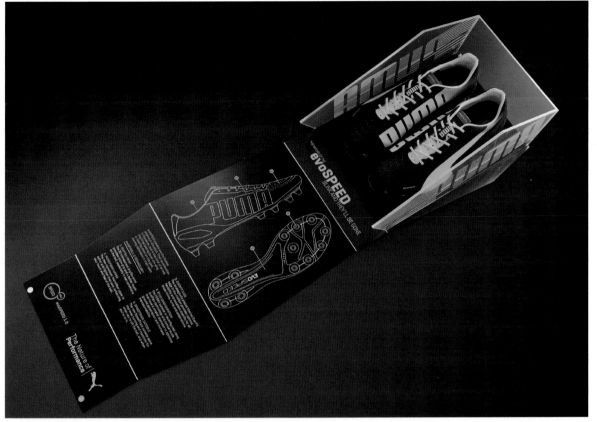

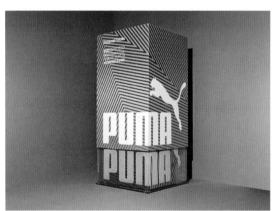
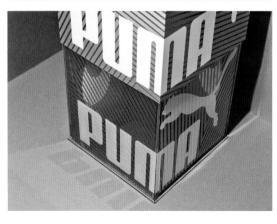
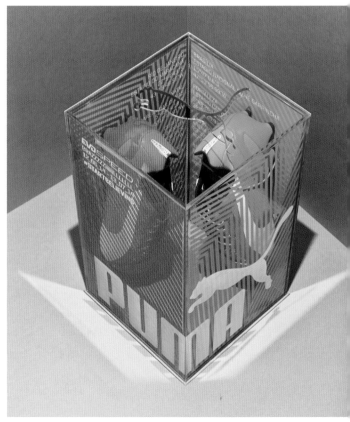

PUMA WORLD CUP 2014
TRICKS COLLECTION

PUMA's mismatched shoes did not fail to catch attention on the 2014 Brazil World Cup football pitch, and limited special packages were created for just a handful of star players. Slide off the outer packaging, a vibrant pink and blue shoebox is revealed. The hinged box has two types of print finishes, matt and gloss. Split open the box vertically, shoes are displayed against a graphic background, featuring the names and coordinates of 12 world cup stadiums in Brazil.

Design: Everyone Associates
Client: PUMA

ADIDAS F50

To launch the new adizero F50 boot, Everyone Associates
created a limited edition of 120 VIP cases for celebrities,
select press and notable bloggers. The process of opening
the box was designed to be as tactile and engaging as
possible to enhance the enjoyment of slowly discovering
its contents, starting with a single personalised boot held
by hidden magnets. The glow edge perspex lid slides off to
reveal a customised football shirt, the miCoach components
and a matching boot in the secret compartment.

Design: Everyone Associates
Client: adidas

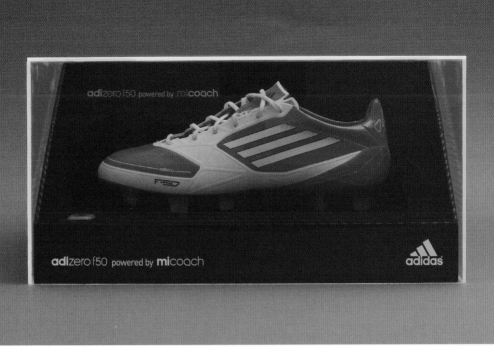

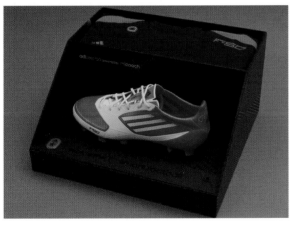

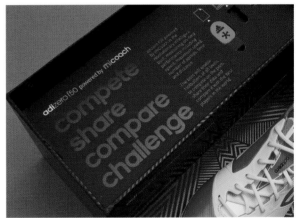

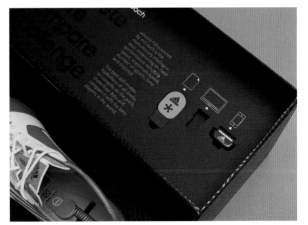

NVRBRKN GLASS SHOEBOX

Entirely made out of glass, the NVRBRKN shoebox was designed for DC Shoes' concept collection. The idea behind the packaging is "Never Broken", a metaphor for athletes' physical and mental power to overcome adversity, most evident in skateboarders. Corresponding to DC's seven-point star logo, only seven shoeboxes were produced and given to seven special individuals, who convey a NVRBRKN point of view in life.

Creative direction & design: Deven Stephens
Production: Soulryde
Photo: Al Quiala
Client: NVRBRKN by DC Shoes

VISION × ILLUSTRATION

VISION ✕ ILLUSTRATION

Vision x Illustration presents art pieces and product packaging created through collaborating with artists. It is about the bold colours, signature patterns and artistic interpretations of product themes that all come together to form into an eye-catching illustration. Mostly featuring designs that merge classic products with wonderfully elaborate skins, these results unfold stories of making the product and engagingly blended visions, adding that a personal introduction to the customer.

THE CUTLERY OF MONOPURI

MONOPURI is a collaborative project experimenting with the possibility of printing effects on various types of materials. Starting from 2012, the project curators have been inviting six pairs of designers and studios each year to create a crossover product based on their expertise. The colourful pattern utensil set from the second edition is the brainchild of AD&D and 501DESIGNSTUDIO. It contains three types of patterns, namely Gradation, Pattern and Surface, all inspired by the elements from the nature.

Art direction: Ren Takaya (AD&D)
Design: AD&D, 501DESIGNSTUDIO
Curation: Shu Hagiwara (MONOPURI)
Client: Magic Touch Japan Co.,Ltd.

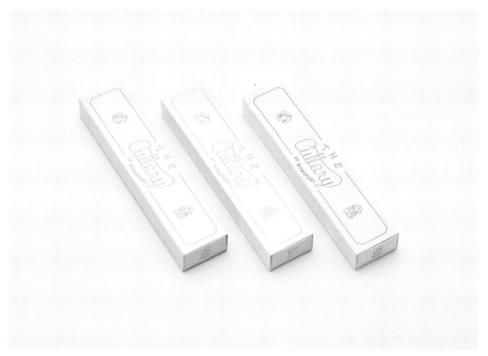

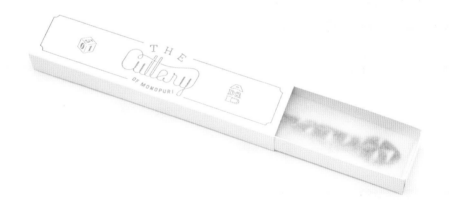

SKETCH

Testing out pens is one of the many pleasures of visiting a stationery store. Creative agency BSENT RESENCE helped transfer such idea onto the visual identity of Thai stationery brand Sketch. Paper boxes and cylinder packagings are filled to the brim with simple patterns drawn by Sketch's range of permanent markers, highlighters and colour pencils. Their colours and strokes are displayed at first glance for easy recognition.

Design: BSENT RESENCE
Client: Sketch

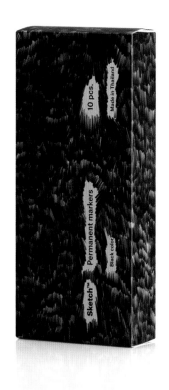

MR. GRAPE & MISS ORANGE

Japanese confectionery company Meiji has been known for tongue-in-cheek packaging designs, one of them being the seasonal release of grape and orange flavours in their XYLISH gum line. Grape and orange are personified by Yamane Yuriko Shigeki's illustrations, each portraying six bizarre alter egos concealed beneath the wrapper. Curiosity and the urge to own a complete set succeeded in attracting sales from the targeted younger generation.

Creative direction: Shumei Takahashi (DENTSU INC. Tokyo)
Art direction: Rintaro Shimohama (DENTSU INC. Tokyo)
Design: Masanori Masuda (DENTSU INC. Tokyo)
Illustration: Yamane Yuriko Shigeki, YAMAMASA
Production: ADBRAIN Inc.
Client: Meiji Co., Ltd

LE CHOCOLAT DES FRANÇAIS

Simply meaning "chocolate made in France", Le chocolat des Français delivers an exclusive French identity that encompasses its brand name, ingredients and packaging. Founders Paul-Henri Masson and Matthieu Escande invited 50 artists to create wrappers for their chocolate bars, with the only criterion of "evoking French". The result is an array of eclectic and fun printed collectibles that delight the eyes.

Creative direction: Paul-Henri Masson, Matthieu Escande
Art direction: Paul-Henri Masson
Illustration: Marie Assénat, Edith Carron, Jean André, Maud Begon, Gaston de Lapoyade, Serge Bloch, Soba, Broll & Prascida, Laura Junger, Paul-Henri Masson
Client: Le chocolat des Français

VELKOPOPOVICKÝ KOZEL
LIMITED EDITION

Velkopopovický Kozel issued a limited edition of their canned beer to commemorate the Old Czechia's heritage and attract new customer base. Ancient traditions and mastership of the Czech brewers are illustrated in detail by simulated wood-etching. Jacketed by a wrapper in the same diamond print, the design reflects on both tradition and craft.

Design: Graphic design studio by Yurko Gutsulyak
Client: Efes Ukraine

ONLY THE BRAVE

Only The Brave is a collaborative project arranged
by Diesel's fragrance line, and only selected creatives
were invited to redesign the identity of the cologne.
John James' work is specifically a photographic
collage referencing two main influences in his life
— naturalist and artist John James Audubon and
the natural world. Electric coloured visual identity
is a merging of many species in nature, such as sea
creature, mammal and carcass. All artworks went on
an exhibition tour that kicked off in New York.

Design: John James
Client: Only The Brave (Diesel, L'Oreal)

VASO DI CULO

Vaso di Culo is a collaboration between Dutch artist Parra and Case Studyo. This porcelain sculpture is a literal cut-out of a woman's body that celebrates the feminine curves from below the waist to the upper thigh area. Black birds are flying around on its cream-coloured skin to look for a resting spot. Limited to an edition of 200, the creation also fulfils functionality of a flower vase.

Design: Case Studyo x Parra

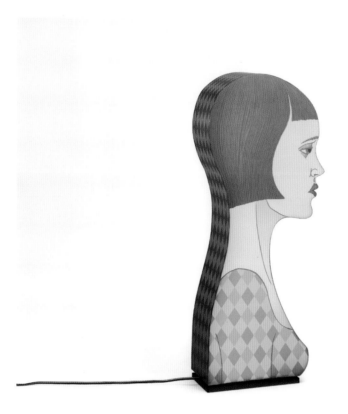

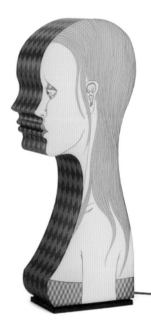

LAMP

Aptly named "LAMP", the 52-centimetre head sculpture stands as a beautiful art piece during the day, and emits atmospheric light at night. Built as an limited art edition for Case Studyo, LAMP reveals Ed Templeton's signature head figure objects with two mopish faces and a chequered profile. The product is shipped in a screenprinted wood box with a numbered certificate signed by the artist.

Design: Case Studyo x Ed Templeton

WOSHI-WOSHI BATH SALTS

A Japanese-inspired bath salt series, Woshi-Woshi features six variations, each with a herb or flower extract for a particular efficacy. With a consistent aesthetic, each packaging leads one to a unique colourful Japanese landscape where its botanic ingredient such as marigold, chamomile or calendula blossoms in the foreground to imply freshness.

Art direction & design: Art. Lebedev Studio
Client: A5 company

KIEHL'S ARTIST COLLABORATION

Specially commissioned for the festive season, Craig & Karl gave Kiehl's classic monotone packaging a makeover with the duo's iconic geometric graphics and colours. To illustrate a joyful Christmas atmosphere, the palette made an allusion to various of colourful Christmas tree ornaments to wrap Kiehl's festive products and gift sets for both male and female customers.

Design: Craig & Karl
Client: Kiehl's

LET IT RAIN

Let It Rain is the result of an artist collaborative project initiated by Harbour City shopping mall in Hong Kong, which has previously featured celebrity artists and designers like Yayoi Kusama and Jeremy Scott. In 2014, the prolific designer duo created a graphic umbrella pattern in two colourways for male and female customers. Other than the cold and warm colour tones, stylish fasteners in cool sunglasses icons also added personality to the umbrellas.

Design: Craig & Karl
Conception: AllRightsReserved
Client: AllRightsReserved, Harbour City

EYES ON THE HORIZON

A successful graphic invasion to the high fashion world, Eyes on the Horizon is German luxury brand MCM's 2013 Spring/Summer collection. Iconic elements that represent the sweaty sunny vacation seasons, such as sunglasses and palm trees, have perked up the collection's leather products like charm keychains and purses. The eye-catching graphics on top of the brand's signature logo pattern Cognac Visetos kindled a fabulous graphic explosion.

Art direction: Craig & Karl
Photo: Alex Sainsbury
Client: MCM

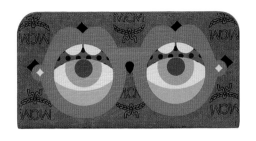

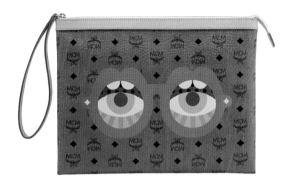

ABSOLUT FUEGO
COLOMBIAN PARTY EDITION

Absolut Fuego was the gold winner of the 2013
Colombian Young Lions Design Competition
initiated by Absolut. Energetic, upbeat and fiercely
colourful, it embodies a shared vibe between
Colombian nightlife and Absolut. Drawing upon
pre-columbian graphics, the wrapper can break into
three parts and rotate to form different characters.
Its thermosensitive print is expected to change
colour as the party heats up.

Design: Emerson Martinez, Diego Almanza
Client: Absolut

COCA-COLA X
STEVEN HARRINGTON

The Coca-Cola x Steven Harrington collaboration
consists of five unique Coca-Cola cans, each illustrated
and designed by the Los Angeles-based artist. Specially
created for the beverage giant's internal team in Spain on
the theme "Share Your Sparkle", each illustrated artwork
represents a facet of Coca-Cola's brand ideology.

Design & illustration: Steven Harrington
Client: Coca-Cola

COLETTE X
STEVEN HARRINGTON

A collaboration between French clothing and beauty
retailer colette and Los Angeles-based artist Steven
Harrington, this 2014 capsule collection features
candles, lighters, T-shirts and a set of postcards.
Harrington's LA-inspired, psychedelic-pop aesthetic
roams the collection and generates chemistry with
colette's Parisian vibe.

Illustration: Steven Harrington
Client: colette Paris

Global merchandise brand Bravado meets Korean creative studio Sticky Monster Lab.

This unique collaboration work will give you a special meaning, you've never seen before.

ROLLING MONSTERS

Korean graphic design studio Sticky Monster Lab (SML) collaborated with music merchandising company Bravado to give Rolling Stone's classic emblem a refreshing look. Celebrating the 50th anniversary of Rolling Stone's debut, Rolling Monsters is a series of raglan T-shirts featuring the legendary band's tongue and lip design tinted in a profusion of patterns. The SML design aimed at becoming a fashion item of rock music scene.

Design: Sticky Monster Lab, Rolling Stones
Client: Universal Music Korea

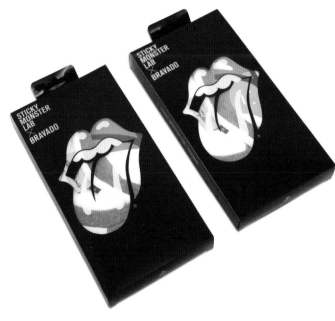

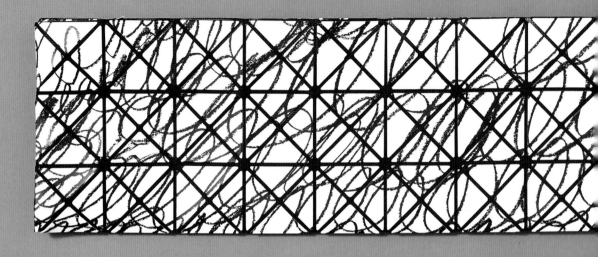
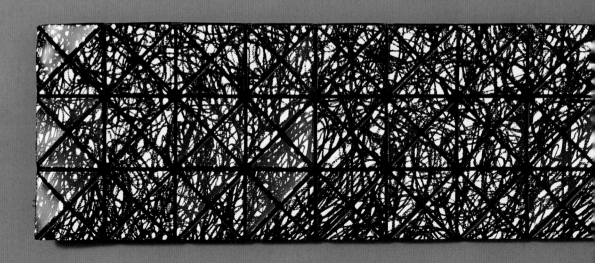

BAO BAO PARK EDITION #03: FREE HAND BAG

Free Hand Bag is one of a series of BAO BAO PARK Edition, a joint project initiated by ISSEY MIYAKE that invites creatives to add fun to the brand's signature BAO BAO bags. Evolving from a clutch, Rikako Nagashima devised four ways to carry the bag in honour of the flexible functionality the design has been known for. Intensity of the freehand pencil print changes as the bag transforms.

Design: Rikako Nagashima x BAO BAO ISSEY MIYAKE

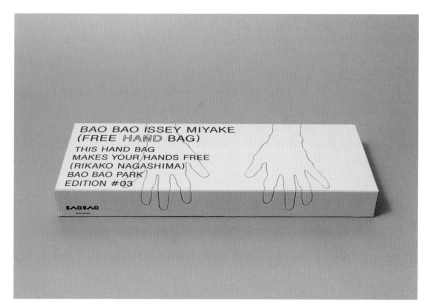

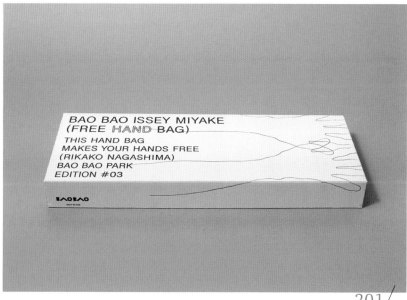

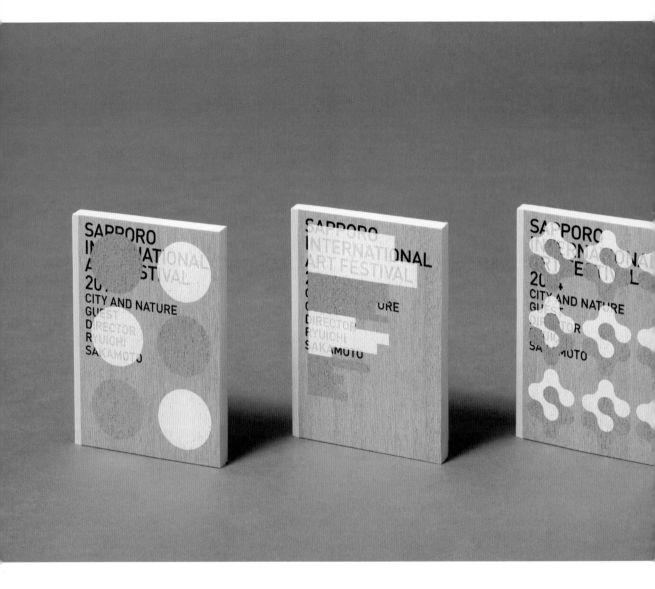

SAPPORO INTERNATIONAL ART FESTIVAL 2014

The 2014 edition of Sapporo International Art Festival (SIAF) was themed "City and Nature". The festival's official designer Rikako Nagashima developed a set of wood-based promotional materials as well as collateral for Forest Symphony. Supported by Louis Vuitton, the Symphony was the festival's featured art project to convert bioelectric potential of trees into music.

Design: Rikako Nagashima
Client: Sapporo International Art Festival

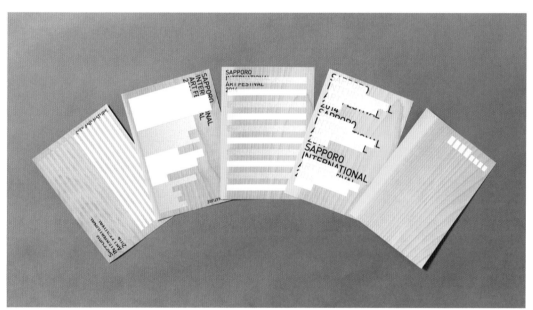

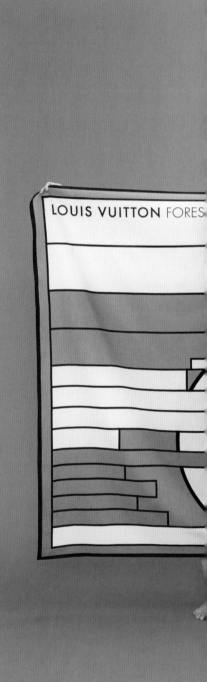

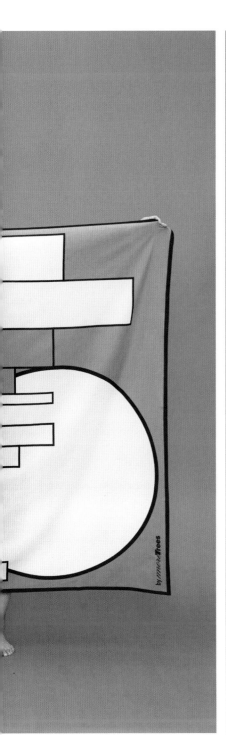

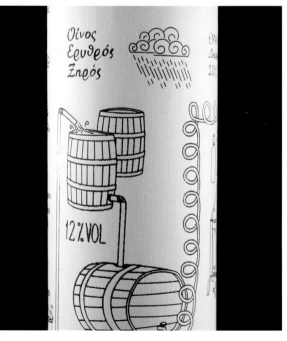

FILIREA GI

"Filirea Gi" (Filiro land) is a small village outside Thessaloniki, where a homemade wine farm produces limited amount of high quality wine every year. Each annual production is accompanied with a brand new packaging. The 2012 edition featured a silkscreened illustration that explained the whole process of production, telling the story from harvesting the produce, bottling to packaging, which all takes place in the small farm. The print is wrapped around the bottles to give out a handmade impression.

Design: Chris Zafeiriadis
Client: Pashalis Zafeiriadis Estate

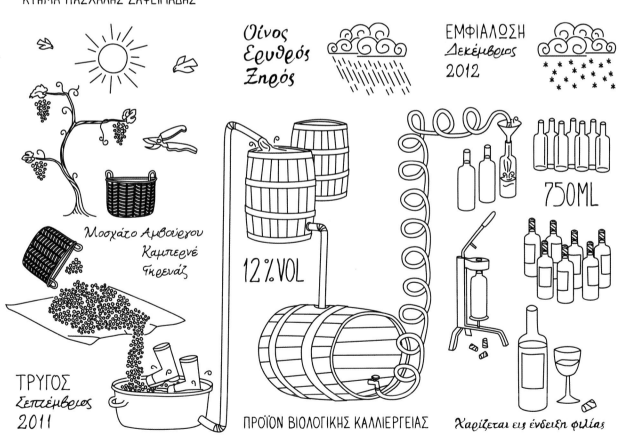

RICE OF SHIROKUMA

Meaning polar bear, Shirokuma is a time-honoured rice label from Niigata, one of Japanese prefectures known for its beautiful, long-lasting snowfall. A lovely rebranding is created for Shirokuma by conjuring polar bears from a grain of rice. The creature's adorable looks add a heartwarming appeal and rejuvenate the century-old brand.

Art direction & design: Frame
Illustration: Masami Sakamoto
Copywriting: Kyoko Nakagawa
Client: Shirokuma co., ltd.

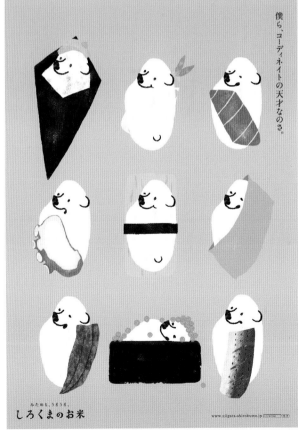

MARINE SWEETS

MATE studio created an unusual packaging for a widely known sweets in Russia, Marine sweets. The packaging connects detailed illustration with minimalistic design. Individual candy wrapper placed inside are folded like a ship, printed in beautiful combination of red and blue. It is also the contrast between the red and blue that enhances the illustration. Drawings include classic icons that we associate with marine, such as anchor, sea creature, waves, and shells.

Design: MATE
Client: Marine Sweets

ALLSORTS PACKAGE DESIGN

Bond Creative Agency has designed a bold and playful packaging for confectionery Cloetta's liquorice line, Allsorts. Boxes are designed to bring the candy's distinctive shapes and colours to the forefront. Contrasting the black and vibrant blocks that resemble the sweets not only makes the packaging stand out from competition but also highlights the new recipe at first sight.

Design: Bond Creative Agency
Photo: Irina Hurme
Client: Cloetta Suomi Oy

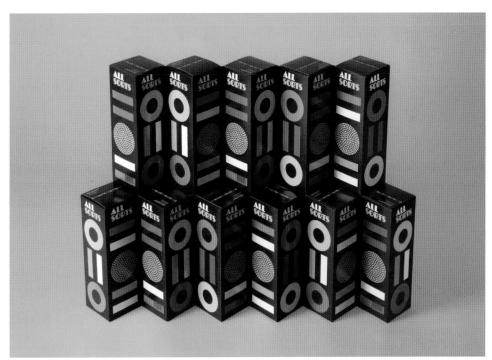

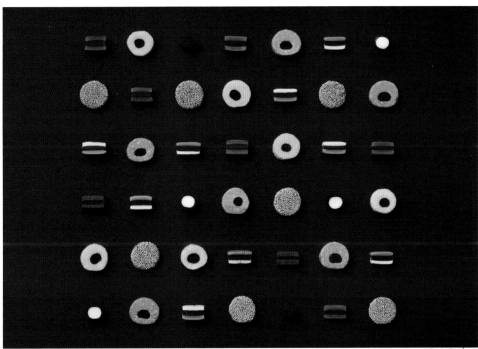

ABSOLUT PITCHERS

Studio C&C created a set of four cocktail pitchers as part of Absolut's Art Pitchers series. The limited art edition features kitsch summer-inspired graphics, extending a lighthearted vibe to parties in any season. Comes with an easy carry handle, the cocktail pitcher doubles as packaging for Absolut bottles besides being a graphic delight by itself. The 70s-inspired still life photo shoot by Oskar Proctor zested up the design and got the party beat going.

Design: Studio C&C
Photo: Oskar Proctor
Client: Absolut

JO MALONE LONDON

Studio C&C created exclusive packagings for five of Jo Malone London's selected fragrances. It was part of Selfridges Beauty Project, a marketing event initiated by the British department store Selfridges. This design visualised the brand's fascinating scents by laying out slices and bits of raw ingredients used in the respective formula, then abstracted the graphics by using print process. Screenprinting sessions were held at Selfridges Birmingham to produce the limited packaging on site, where customers could pick their own colour palette.

Design: Studio C&C
Photo: BBDE
Client: Jo Malone London

LORETTA SECRET GARDEN

Relating to the little girl in each woman's heart, illustrative packaging
has always been Loretta's signature marketing approach. The Japanese
cosmetic label worked with illustrator Noriko Yamaguchi to develop
packaging and brochure for their organic skincare line, Secret Garden. A
dreamy, thriving garden is created, telling the story of plants that brings
happiness to people.

Creative direction: Aya Obayashi (beauty experience Inc.)
Design: Sayaka Koda (Nippon Design Center, Inc.)
Illustration: Noriko Yamaguchi
Copywriting: Naho Yoshioka (Nippon Design Center, Inc.)
Client: MoltoBene INC.

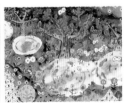

KOMORI MANGO

Branding for mangoes from Komori Farm is inspired by the fruit's vibrant red colour and its lush green leaves that grow under the bright sun. Clustering around the sun-resembling logo, soft watercolour foliage with mangoes among them flourishes on boxes, catalogues and wrapping paper, which doubles up as posters.

Art direction & design: Kimiko Sekido (DENTSU INC.)
Client: KOMORI FARM

愛情をかけた分だけ
マンゴーはおいしくなる。

マンゴーはとても手間のかかる果実です。
樹や果実の生育に合わせて
ハウス内の温度をこまかく調整することはもちろん
葉や枝で日光が遮られないように
ひとつひとつ果実の向きや高さを調整しなくては
おいしいマンゴーはできません。
小森農園のある宮崎県は、日照時間が全国トップクラス。
マンゴー栽培には最適の場所です。
太陽に恵まれたこの場所で、丁寧に育てた特別なマンゴーを
お召し上がりください。

宮崎県産完熟マンゴー　小森農園

KIRIN YAWARAKA TENNENSUI

SAGA INC. brought about a set of decorative patterns for Kirin's natural mineral water packaging. The objective is to create a design that harmonises and decks the interior. Packed in a carton of six, each bottle in the set wears a unique aquatic pattern that together tells the story of how quality mineral water originates from nature.

Design: SAGA INC.
Client: Kirin Beverage Company, Limited

ARTISAN TRUFFLE PACKAGING

This packaging is designed to deliver a complete experience of Costello & Hellerstein's sublime chocolate truffle — reflecting creativity, technical brilliance and a collaborative approach. The wordmark is crisp and clean in contrast to the free flowing marble pattern in the background. The effect was achieved by swirling ink on the surface of water, and overlaying cartridge paper. The package fittingly expresses the brand identity and the product within the package.

Design: Robot Food
Client: Costello & Hellerstein

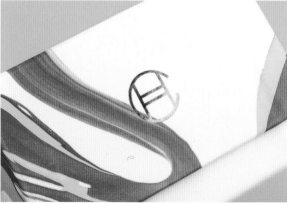

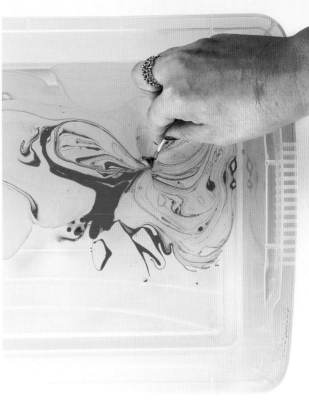

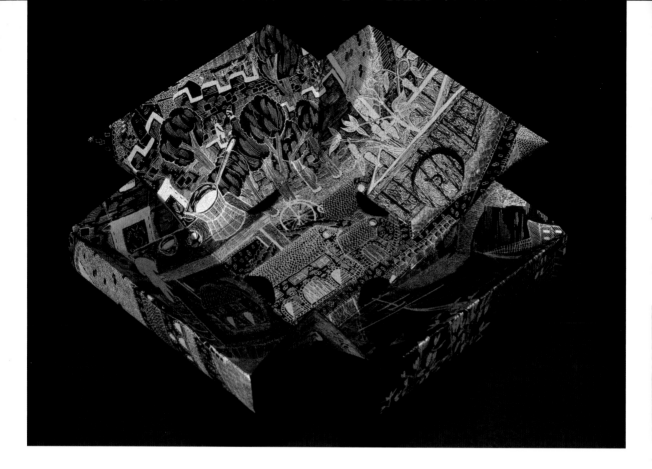

EACH HIS OWN

Each His Own pepper cookies is Croatian brand Paprenjak's flagship product. Designed to unfold as a plate when opened, its limited souvenir packaging is a premium box with intricate illustration that portrays the landscape of old Croatia and how pepper used to be transported from sea to land. Imaginative features of sea monster and the witch Magda selling Paprenjak in town square can also be found in the illustration.

Creative direction & design: Ivorin Vrkaš (Rational International)
Illustration: Marina Milanović
Consultation: Ruđer Novak-Mikulić
Client: Paprenjak

MARTINI ASTI CHRISTMAS PACKAGING CONCEPT

Martini Art Club hosted a package design competition, where artists submitted their own ideas for Martini Bianco and Martini Asti bottles. Alexandra Istratova's design features two distinctively patterned covers. One of which is a cool-toned pattern that resembles leafs, and the other has a warm-toned pattern with integrating swirls. Whether it is a package for a sophisticated event or a night out in the city, the attaching masks will not only match but also elevate the mood of the party.

Design: Alexandra Istratova
Client: Martini Art Club

ABSOLUT ORIGINALITY

Absolut has always been experimental about new techniques and daring concepts that realise the next limited edition. Absolut Originality is one of the brand's creative endeavour in 2013, celebrating a triumph over fusing traditional Swedish glass craft into modern production. A drop of cobalt blue is infused into every molten glass just before they go into the mould, resulting in unique stripes gracing each of the four million bottles distributed worldwide.

Design: Absolut
© The Absolut Company AB.
Used under permission from The Absolut Company AB.

SVEDKA SUMMER EDITION

Every year Studio Established designs a special packaging
for vodka brand SVEDKA to celebrate Independence Day
in the States. Centring on the significance of this day,
these designs are conceived based on the Stars and Stripes
theme. From 2012 to 2014, three shrink-wrappings for the
brand's standard 750ml bottle were introduced, each with a
distinctive graphical twist.

Creative direction: Established
Client: SVEDKA

SVEDKA TITANIUM

Endorsed by the famed SVEDKA robot, SVEDKA Titanium is the Swedish vodka brand's tribute to 2012 holiday season. Adhering to its tagline "Party Like it's 2033", studio Established conjured a mind-blowing optical illusion that reflects the intoxicating effect of vodka as well as wild fun at parties. The pattern is shrink-wrapped around the bottle. The amusing effect naturally makes the product pop on the shelf.

Creative direction & design: Established
Client: SVEDKA

CAMPBELL'S WARHOL SOUP CANS

To coincide with the 50th anniversary of Andy Warhol's first iconic soup can paintings, Campbell Soup commissioned four limited edition tomato soup can designs, with exclusive placement at Target stores. The artworks paid tribute to Warhol with vivid colours straight from his signature palette, a dynamic logo and narrative detailing his impact on pop art. The artist's portrait, signature and four famous quotes gave an authentic voice to these mini masterpieces.

Design: DDW, BBDO New York
Client: Campbell Soup Company
Special credits: Deutsch Design Works (DDW), Campbell Soup Company, The Andy Warhol Foundation for the Visual Arts, Inc., Jilliann Smith, Weber Shandwick
© 2015 The Andy Warhol Foundation for the Visual Arts, Inc. / Artists Rights Society (ARS), New York. Trademarks licensed and all rights reserved by Campbell Soup Company.

FOOD FOR THOUGHT

Designer Maria Mordvintseva-Keeler thinks that some books are not just books but "food for thought". With this in mind, she created a unique canned packaging for three of her hand-picked classics with edible names. In perfect "serving size", each novel has humorous and sarcastic "nutrition facts" illustrated on the can without revealing the plot of the novel.

Design: Maria Mordvintseva-Keeler

JOHNNIE WALKER
1910 SPECIAL EDITION

Produced to commemorate the opening of Johnnie
Walker House in Shanghai, this special edition
embodies a Chris Martin illustration about Johnnie
Walker's Scotland-Shanghai voyage in 1910. British
studio LOVE transferred the illustration from a wall-
length copper artwork at the Shanghai venue to
willow pattern porcelain bottles limited to
a production of 1000.

Design: LOVE
Illustration: Chris Martin
Production: Wade Ceramics
Client: Diageo

JOHNNIE WALKER
2012 CHINESE NEW YEAR
LIMITED PACKAGING

To celebrate 2012 Chinese New Year, Johnnie
Walker launched a special packaging for their
well-known whiskies. Appropriately fitted to
the year of the dragon, illustrator Chris Martin
incorporated these legendary creatures into his
artwork for Red, Black and Gold Labels limited
edition cases. Detailed drawing in festive gold
colour tells the story of Johnnie Walker bringing
his whisky to an enchanted China.

Illustration: Chris Martin
Client: Johnnie Walker, China

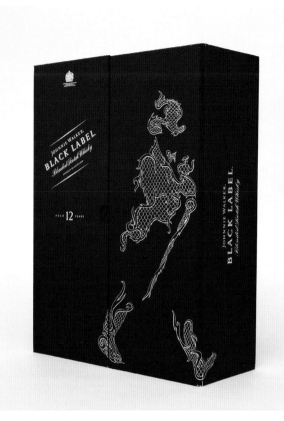

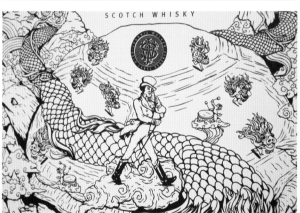

URBAN EXPRESSIONS

Russian Bear Vodka held a design competition,
themed "Urban Expressions". In crisp black and
gold, Hylton Warburton's winning entry portraits
local music, slang, fashion and the overall vibrant
nightlife and culture in South Africa where he's
from. A limited edition of 100 bottles were made.

Design: Hylton Warburton
Client: Russian Bear Vodka

BIOGRAPHY

A - C

ABSOLUT

"ABSOLUT® VODKA. ABSOLUT COUNTRY OF SWEDEN VODKA & LOGO, ABSOLUT, ABSOLUT BOTTLE DESIGN AND ABSOLUT CALLIGRAPHY ARE TRADEMARKS OWNED BY THE ABSOLUT COMPANY AB." Absolut is the world's fifth largest premium spirits brand. Every bottle is produced in Åhus, southern Sweden. The Absolut Company has the worldwide responsibility for the production, innovation and strategic marketing of Absolut, Malibu, Kahlúa, Wyborowa, Luksusowa and Frïs. The Absolut Company is a part of Pernod Ricard, which holds one of the most prestigious brand portfolios in the sector.
P. 230-233

AD&D

Founded by Ren Takaya in 2011, AD&D focuses on graphic design and art direction for signs and logos, campaign shops, and restaurants. Takaya was born in Sendai, Miyagi prefecture in 1976 and graduated in sculpture from Tohoku University of Art & Design in 1999. He joined good design company before establishing AD&D and has been receiving numerous awards including Grand Prix in 2014.
P. 164-165

ALIEN AND MONKEY

A design workshop created by Daishu Ma, a writer and illustrator, as well as Marc Nicolau, an industrial designer.
P. 110-111

ALMANZA, DIEGO

Besides working as a creative director at SANCHO BBDO, the Colombian graphic and furniture designer is also a freelance photographer.
P. 190-191

ALPHA245

A global communications network that generates creative solutions to accelerate business growth for clients. Their approach is rooted in clarity of purpose, creative connections, and open collaboration. Diverse creative minds expand expertise, insight and creativity to disrupt convention, and enhance local business solutions. As the Leo Burnett Group, Alpha245 aims to be the world's best creator of ideas that move people.
P. 033

ALT GROUP

Based in Auckland, New Zealand, the multidisciplinary design studio founded in 2000 by Ben Corban and Dean Poole is now a core team of 25 people with diversified backgrounds and experiences in strategy, content design, graphic design, interactive design and three-dimensional design. Alt has been recognised in numerous international awards including ADC, AIGA, AGDA, Cannes Lions, The One Show, Red Dot and TDC.
P. 020-021, 120-121

ARBUZOV, MAKSIM

Started by drawing with a pencil on walls, Arbuzov later went for spray paints on cars, houses and city streets. He then worked for a student newspaper as an illustrator while studying tourist management which has never turned into the designer's career.
P. 124-125

ART. LEBEDEV STUDIO

A privately held company offering advanced industrial, graphic, web, and interface design. Working to find the most simple, elegant, and convenient solution to any problem without losing the purpose. Based in Moscow with two other offices in Kiev and New York, the studio in principle doesn't work with private persons, political parties, religious organisations, jerk-offs, and those whose views conflict with theirs.
P. 180-181

BAILEY LAUERMAN

An independent advertising agency headquartered in Omaha, USA, a place that Lauerman sees it free to explore new ways of thinking, giving them an independent streak as wide as the plains. The team is not held hostage to any one perspective, medium, category or holding company. The only goal is the relentless pursuit of positively impacting the destiny of the brands they serve.
P. 068-069

BARB TEAM

The collaboration of single-minded multidisciplinary designers from Moscow, Russia. Barb Team includes Roma Lazarev, Sergei Anenko, Alexandr Romanov, Anton Khabarov, and Elena Khoroshiltseva.
P. 102-105

BARTEAUX, TRAVIS, NIKE INC.

With experience spanning advertising, identity, print, web and environments, Barteaux is a multidisciplinary designer who has worked in-house at Nike, Saturdays NYC, Billy Reid, as well as agencies like Mother New York, VSA and High Tide.
P. 152-153

BELLO, CHI-CHI

Bello is a senior art director and designer living in San Francisco, California, USA.
P. 078-079

BEND

Operating since 2005, Bend is a flexible creative team which strongly believes in good, functional design as much as in effective art direction. Their interests lie in the wide spectrum of visual communication but they specifically work in the fields of graphic design, illustration, typography screenprinting and interactive media.
P. 132-133

BIGGERS, SHAR

Graduate of Portfolio Center in Atlanta, USA. with a business background, Biggers takes pride in her creative versatility which applies to a number of media environments from design, branding and advertising. The designer is also a dreamer and accomplished craftsperson.
P. 134-135

BOLIMOND, CONSTANTIN

Born in Belarus, Constantin (Osipov) Bolimond studied communication design in Vitebsk State Technological University. The 26-year-old designer has since worked in the branch of visual communication and he mainly focuses on identity-, packaging-, and motion design.
P. 050-051

BOND CREATIVE AGENCY

A brand-driven creative agency with a craftsman attitude. Bond creates brands for new businesses and revolutionises existing ones for growth. Bond brings together talents from different fields to create cross-disciplinary solutions within identity, digital, retail, spatial, packaging and product design.
P. 212-213

BSENT RESENCE

A creative agency founded in 2014 as a research and development department of jewellery brand ABSENT PRESENCE.
P. 166-167

CASE STUDYO

Launched by TOYKYO, Case Studyo is a producer and publisher collaborating contemporary artists from a broad spectrum. The new artist collaboration initiative with dedicated focus and ambition aims to publish limited and open edition objects and artworks with a striking visual language.
P. 058-061, 176-179

CAVIAR DIGITAL

Striving to deliver meaningful and unique experiences seamlessly across all digital media, Caviar Digital's heritage is beautiful storytelling breaking out of the digital sphere and ultimately creating a piece that you can actually touch.

CLARA VON ZWEIGBERGK DESIGN

Since 1995, Zweigbergk has worked as a graphic designer and art director in Stockholm, Los Angeles and Milano. Zweigbergk is now running her own studio in Stockholm combining graphic design, art direction and product design.

CRAIG & KARL

Based in two cities across the ocean, Craig & Karl is a design duo formed by Craig Redman and Karl Maier. Signature in the colourful, geometric pattern face, they work on projects in various medium including murals, sculpture, illustration, installation, interface design, typography, iconography, fabric patterns and print.

D - F

DDW

Formerly Deutsch Design Works, the award winning full-service branding agency headquartered in San Francisco. Led by Co-CEO/Creative Director Ross Patrick and Co-CEO/Director, Client Services Theresa Scripps, DDW's hybrid background includes package design, advertising, corporate identity, sustainable design practices, digital, retail and strategic branding. The team loves to discover and make cool things. They work to solve the most challenging problems with unique, fun, smart and engaging ideas. Strategic insights and big brand experiences gives the team an advantage when it comes to understanding the current retail and digital landscapes. They approach work with an extreme sense of dedication to excellence.

DENTSU INC.

DENTSU INC. has a diverse client portfolio. The company handles advertising campaigns of blue-chip companies, and major global clients have chosen them in return as a partner in the Japanese market. Graduated from Tama Art University Department of Graphic Design, Kimiko Sekido is now an art director at DENTSU.

DENTSU INC. TOKYO

Under the global network of DENTSU INC., which maintains the top share in the Japanese advertising market, the Tokyo office offers all-round services from planning to production, implementation, and management for advertising campaigns which also includes market research, consultation and public relationship. Their media of work covers printing and publishing, digital, and events.

ESTABLISHED

Set up in 2007 by designer Sam O'Donahue and partner Becky Jones, Established is a full service boutique agency offering architectural, graphic and product design under one roof with a focus on an intimate creative partnership between client and studio. Since its formation, Established has won numerous international design awards and has been widely published in the world's leading design books and periodicals.

EVERYONE ASSOCIATES

A creative studio of good listeners and hard thinkers striving to create innovative and effective visual communication. They tell creative stories for brands, in pictures, words, objects and environments. Stories designed to cut through the noise and change how people think and behave. Making a difference to clients' business is what makes them tick, whether they are big or small, global or local.

FAMILIAR STUDIO

Founded in 2012 by Ian Crowther, Carl Williamson and Keith Mancuso whose diversed backgrounds includes activism, art, politics, publishing and the non-profit world, Familiar works to create interactive tools, visual systems and technological solutions for a wide range of clients.

FONSECA, DIEGO

Fonseca is a Brooklyn-based art director with a passion in anything nerdy. He likes to spend his free time with fun projects and pop culture cross overs.

FRAME

Without being bound by preconceived ideas while having constant consideration on new mechanism and gimmicks, Frame works to create new values for communication design via corporate identity and visual identity planning, product and brand development, and advertising production.

G - I

GRANTIPO

Based in Madrid, Spain, Grantipo is a design studio specialising in branding and packaging.

GRAPHIC DESIGN STUDIO BY YURKO GUTSULYAK

Born in 1979 in Ukraine with an academic background in economics, Gutsulyak began his design career when he moved to Kiev in 2001. Founded the studio in 2005, Gutsulyak's work has been widely published and exhibited internationally including France, Poland and China. The winner of more than 30 international awards was elected as the first president of Art Directors Club Ukraine in 2010.

GRAPHICAL HOUSE

A design consultancy located in Glasgow, UK producing thoughtful and crafted work across all applications including digital, analogue and environmental.

HARRINGTON, STEVEN

Cited as the leader of a contemporary Californian psychedelic-pop aesthetic, the Los Angeles-based artist and designer is best known for his bright, iconic style. Embracing a multimedia approach, Harrington's portfolio includes large-scale installations made of plaster and stone, handscreened prints, limited-edition books, skateboards, and sculptures. Alongside his personal work, Harrington co-founded both the acclaimed design agency National Forest and pop-art brand "You&I".

HELLSTEN, SAANA

The Finnish multidisciplinary designer has a diverse background ranging from graphic and packaging design to fashion art direction and marketing. Her work has been published in several international books and she has won a number of awards and scholarships. She completed MS in Package Design at Pratt Institute and BA in Design at Lahti Institute of Design. Being a Scandinavian, she appreciates nature, sustainability and equality. With experience living and studying in Berlin and Italy, Hellsten is now based in New York, USA.

ISTRATOVA, ALEXANDRA

An artist, designer and art-director, Istratova is also a member of Designers' Union of Russia and just a cute girl. Studied graphics at Tulskiy State University and visual communication at British Higher School of Art and Design, Istratova is the winner and finalists of different international and Russian festivals of design such as Pentawards, ADCR, ADCE, Idea, White Square, etc.
P. 228-229

J - L

JAMES, JOHN

Born and bred in Canada, the art director spends most of his days hunting brilliant minds and unyieldingly passionate hearts that help shaping his aesthetic and who he is today. He has worked for clients big and small including Google ATAP, Nike, T-Mobile, the NBA, Corbis, Mercedes Benz USA, and Diesel, among others.
P. 174-175

KATO, KAORI

A Tokyo-based graphic designer and art director. Kato graduated from the Tohoku University of Art & Design in 2009 and got her first edition of "Skirt-Flipping Calendar" published and sold all over Japan in 2012.
P. 022-023

KEIKO AKATSUKA & ASSOCIATES

A graphic and packaging design studio in Tokyo, Japan winning the 2015 Pentawards Gold and Japan Package Design Competition 2015 Selected, to name a few.
P. 086-087

KOKESHI MATCH TEAM

Consisting Kumi Hirasaka, Shinsuke Nishiumi, Akiko Yamada, the team sells matches with faces made in bulk in a factory. Other than these matches, they have one-of-a-kind matches made as an art for the exhibition in 2011.
P. 026-029

KOLLE REBBE GMBH

Kolle Rebbe is one of the top creative agencies in Germany that conduct business with entrepreneurial intelligence on their cooperative interdisciplinary approach, transcend boundaries of culture and media and believes in common sense.
P. 106-107

LATONA MARKETING INC.

A Japanese design office adept with marketing strength founded by designer Kazuaki Kawahara in 2008. Latona conducts precise marketing in advance, enabling the team to correctly pose problems, derive the correct solutions, and reflect these in their designs. The ultimate objective is to increase sales and enhance the management of client companies.
P. 034-035

LAVERNIA & CIENFUEGOS DESIGN

A multidisciplinary design studio based in Valencia, Spain that specialises in graphic, product and packaging design, Lavernia & Cienfuegos has a strong international reputation with clients in Japan, China, Russia, Belgium, UK, Brazil, Switzerland and Spain.
P. 090-091

LE CHOCOLAT DES FRANÇAIS

Founded by Paul-Henri Masson and Matthieu Escande, Le chocolat des Français is a new chocolate brand providing hand-crafted chocolate in colourful packagings.
P. 170-171

LELIĆ, DENIS

Slovenian creative with a great passion for typography, brands and simplicity. Currently working out of Ljubljana as a design director for Alkemia, Lelić produces a wide range of work stretching from web to illustration and branding for clients from all over the world.
P. 044-045

LEWIS, GWYN M.

Born and raised in the San Francisco Bay Area, USA, the visual designer has been freelancing to express her own style besides a full-time design job. Graduated with a masters degree in graphic gesign in 2014, Lewis took an elective in package design and it became her passion ever since.
P. 070-071

LIN, YU-HENG

Born in 1994 in Taiwan, Lin is a current student of Commercial Design and Applied Foreign Languages at National Taiwan University of Science and Technology, where she studies to specialise in graphic, editorial and packaging design.
P. 108-109

LO SIENTO

Founded by Borja Martinez in Barcelona in 2005, Lo Siento is specially interested in taking over identity projects as a whole. The main feature of its work is its physical and material approach to the graphics solutions, resulting in a field where graphics and industrial go hand by hand. In 2010, the studio was awarded by the FAD (Fomento de las Artes Decorativas) with the Grand Laus award for the project: EMPO.
P. 046-047

LOVE

LOVE is a design studio, an ad agency, and an innovation kitchen with creative thinkers, strategy authors, experience creators, content makers, and specialists working together to create compelling ideas for global brands. Recent LOVE brands include Virgin Atlantic, Haig Club, Guinness, Adidas, and the cover design of the world's fastest selling album for One Direction.
P. 240-241

M - O

MARCH DESIGN STUDIO

A product development company founded in 2012 and currently located in Vilnius, the capital of Lithuania, MARCH invents its own products and collaborates with other creative folks with an interest ranging from life saving reflectors to handmade lollipops.
P. 072-073

MARTIN, CHRIS

An award-winning illustrator and designer based in London, with over five years' experience working in advertising, editorial, animation and fashion. Martin's clients include Johnnie Walker, The Guardian, Ray-Ban, Yazoo, Nokia and Nike.
P. 242-243

MARTINEZ, EMERSON

Art director at Leo Burnett Colombia, Martinez is also an illustrator for many publishers and dreams of becoming a great designer.
P. 190-191

MATE

Founded in Saint-Petersburg in 2014 by two designers Alice Macarova and Dmitry Terehov, MATE works to focus on branding, illustration and packaging design.
P. 210-211

MENTXAKA, TXABER

Born in Bilbao, Spain and studied graphic design from 1988 to 1992, Mentxaka has developed an intense professional life as a graphic designer working for different agencies and a large range of clients.
P. 036-037

MORDVINTSEVA-KEELER, MARIA

A multidisciplinary designer with a focus on packaging, branding and custom typography. Originally from Moscow, Russia, Mordvintseva-Keeler is based now in San Diego, USA. In her work she aspires to create designs that push the boundaries, have a personality and a touch of humour.
P. 238-239

MORUBA

Formed by Daniel Morales and Javier Euba, Moruba's projects often transcend the initial objectives and end up being recognised by national and international awards, as well as published in books and magazines.
P. 038-040

NAGASHIMA, RIKAKO

Born in 1980, the art director and designer graduated in Visual Communication Design at Musashino Art Universaty in 2003. Nagashima's work covers a wide range of areas including signage design, advertising, branding, spatial design, product design, film production, and music video. The award-winner was also a member of ADC who started her own brand "Human_ Nature" in 2013.
P. 200-205

NENDO

Founded by architect Oki Sato in 2002 in Tokyo, nendo holds its goal of bringing small surprises to people through multidisciplinary practices of different media including architecture, interiors, furniture, industrial products and graphic design. Sato was chosen by Newsweek magazine as one of The 100 Most Respected Japanese and won major awards include "Designer of the Year" by Wallpaper magazine and "Guest of Honor" from Stockholm Furniture Fair.
P. 008-009, 016-019, 092-095

NERETIN, STAS

Born in the city of Voronezh, Russia, Neretin first graduated in graphic design at Architecture University of Voronezh and later from the British Higher School of Art and Design. Doing graphic design for over twelve years, the chief designer is now working in the Restaurant of Association in Moscow.
P. 130-0131

NIPPON DESIGN CENTER, INC.

Established in 1959, Nippon Design Center, Inc. is Japan's largest and most creative corporation, with a staff of 237 providing high quality services in graphics, packages, corporate identities, images and WEBs. Their clients includes Toyota Motor Corporation, Ryohin Keikaku Ltd. and Nikon Corporation.
P. 014-015, 218-219

OGILVY & MATHER BOGOTÁ

One of the largest marketing and communications companies in the world, O&M was named the Cannes Lions Network of the Year for 2012, 2013, and 2014; and the EFFIEs World's Most Effective Agency Network for 2012 and 2013. O&M services Fortune Global 500 companies as well as local businesses through its network of more than 500 offices in 126 countries.
P. 116-119

OLSEN & OSLO

The work name of Charlotte Olsen, a Norwegian branding and packaging designer known for a quirky design style with strong, bold colours and shapes. Simply going for the obvious classic design principles, Olsen believes in taking an idea and trying to do the opposite to challenge each product and design category with the goal to achieve more innovative results each time.
P. 030-031

P - S

POULSEN, JUSTIN

The passion in breaking things, painting, building and making has led to specialisation in conceptual still life and photo illustration. Poulsen's photography has been acknowledged worldwide by publications and award shows such as the International Photography Awards and PX3 Prix de la Photographie Paris.
P. 052-053

QUITLLET, EUGENI

Graduated from the "La Llotja" art school in Barcelona, the Catalan designer synthesises drawing and sculpture creating objects that masters the fullness and emptiness to reveal elegant silhouettes hidden in the material. Quitllet's creative vocabulary exceeds the simple relationship between function and style to introduce new objects to the current design scene, never lacking in communicative enthusiasm.
P. 041

RAMONES, RYAN

A young graphic designer and art director from Rotorua, New Zealand. Romanes has worked in various places before relocating to Melbourne, Australia. His work has been celebrated in the New Zealand Best Awards and featured by numerous publishers like Domus, Ignant, and The Dieline. He now works with local and international clients of all business backgrounds for graphic design and image making projects.
P. 140-143

ROBOT FOOD

Specialising in branding, packaging, innovation, creative spaces and web, Robot Food is a Leeds-based strategic creative agency founded in 2009. Clients include plucky challengers through to global FMCG players and 'bold, brave and beautiful' is their blueprint for design.
P. 080-083, 224-225

SAGA INC.

Formerly named choudesign, SAGA was founded in 1995 offering a wide range of services including branding, visual and corporate identity, web design, package design, advertising and art direction. Led by Kota Sagae, SAGA has received numerous awards including the 2015 Japan Package Design Awards.
P. 222-223

SAGMEISTER & WALSH

Sagmeister & Walsh is a New York City-based design firm that creates identities, commercials, websites, apps, films, books and objects for clients, audiences and the team itself.
P. 062-063, 122-123

SAITO, TOMONORI

The son of a Buddhist monk who grew up in a temple, Saito is an art director graduated from Musashino Art University, Japan and started his professional career at DENTSU INC. where he specialises in artistic supervision of commercial advertisements and consultancy of corporate brand management.
P. 012-013

SCHNECK, SHANE

In 2010, American industrial designer Schneck founded Office for Design in Stockholm, Sweden, working to challenge industry standards and discover new solutions to contemporary living. The studio has been honoured by Wallpaper magazine, International Forum Hannover, Guldstolen as well as being collected by global giants such as Google and Skype.
P. 126-127

SCHULZ, JOHANNES

A designer and visual artist based in Hamburg, Germany, Schulz's current workfields are 3D-, CGI-, packaging- and corporate design. There is an outstanding aesthetic quality and strong artistic approach characterising his implementation of work.
P. 048-049

SPEC!AL EDiTION

ARTIST COLLABORATIONS
ON PACKAGING DESIGN

First published and distributed by
viction workshop ltd

viction:ary

viction workshop ltd
Unit C, 7/F, Seabright Plaza, 9-23 Shell Street,
North Point, Hong Kong
Url: www.victionary.com
Email: we@victionary.com
 www.facebook.com/victionworkshop
 www.twitter.com/victionary_
 www.weibo.com/victionary

Edited and produced by viction:ary

Concept & art direction by Victor Cheung
Book design by viction workshop ltd

ISBN 978-988-12227-2-5
Printed and bound in China

ACKNOWLEDGEMENTS

We would like to thank all the designers and companies who
have involved in the production of this book. This project
would not have been accomplished without their significant
contribution to the compilation of this book. We would also
like to express our gratitude to all the producers for their
invaluable opinions and assistance throughout this entire project.
The successful completion also owes a great deal to many
professionals in the creative industry who have given us precious
insights and comments. And to the many others whose names
are not credited but have made specific input in this book, we
thank you for your continuous support the whole time.

FUTURE EDITIONS

If you wish to participate in viction:ary's future projects and
publications, please send your website or portfolio to
submit@victionary.com